P9-DIB-875

## Praise for *The Genesis Machine*

"*The Genesis Machine* is a brilliant pairing of two visionaries who offer us a comprehensive take on making a better world through biology."

—JANE METCALFE, cofounder of *Wired* and CEO of NEO.LIFE

"*The Genesis Machine* is a very readable story about how the DNA world is shifting from reading the genetic code to writing and editing it. Amy Webb and Andrew Hessel then take the reader on a journey of possible world changing events that could result from this new technology."

—J. CRAIG VENTER, PHD, author of *Life at the Speed of Life: From the Double Helix to the Dawn of Digital* and CEO of JCVI

"This spectacular and highly accessible book clearly and thoughtfully examines the most important revolution of our lives—and of life itself. Understanding how we and future generations will use the tools of synthetic biology to transform the worlds inside and around us is essential to being an informed and empowered person and citizen in the twenty-first century. *The Genesis Machine* is a guide to exactly that and a must-read book."

—JAMIE METZL, member of WHO expert committee on human genome editing and author of *Hacking Darwin: Genetic Engineering and the Future of Humanity*

"You may not realize it yet, but your life—and all of life itself—is about to change. From programmable genes to designer medicines, synthetic biology is going to transform everything. *The Genesis Machine* is a surprisingly intimate, incisive, and readable guide to the opportunities, risks, and moral dilemmas of the brave new world ahead."

—STEVEN STROGATZ, Cornell University, author of *Infinite Powers*

"If future technologies arrive gradually and then seemingly all at once, then the biotech-driven future is poised to arrive in ways that are far beyond the reach of our imaginations and at the same time knocking at our doors right now. Amy Webb and Andrew Hessel offer an essential guide to understanding biotech frontiers, and they outline important questions and approaches to consider now. An essential book for business leaders."

—BETH COMSTOCK, author of *Imagine It Forward* and former vice chair, GE

"*The Genesis Machine* is a tour de force! Amy Webb and Andrew Hessel masterfully reveal the emerging network of forces—people, labs, computer systems, government agencies, and businesses—that will drive humanity's next great transformation. Their fascinating (and frightening) conclusions—that the human ecosystem can actually become programmed—will touch every facet of our lives in the future. This brilliant work is an absolute must-read for national security professionals and defense planners who need to understand the complex dynamics at play in the future competition for bio-hegemony."

—Dr. Jake Sotiriadis, chief futurist, United States Air Force

"We can now program biological systems like we program computers, with artificial intelligence and machine learning accelerating the speed of innovation and applications of synthetic biology. In an accessible and fascinating narrative, *The Genesis Machine* lays out a roadmap for this interdisciplinary field of synthetic biology that is forever reshaping life as we know it."

—Rana el Kaliouby, author of *Girl Decoded: A Scientist's Quest to Reclaim Our Humanity by Bringing Emotional Intelligence to Technology* and deputy CEO of Smart Eye

"Are latest innovations in synthetic biology simply a miracle that ends a crisis or a breakthrough to an entirely new way of living? That's the question futurist Amy Webb and microbiologist Andrew Hessel reveal for us with this fascinating book. The history of the world is a history of unintended consequences, for better and for worse, and Webb and Hessel capture the coming fusion of tech and biology in vivid detail."

—Ian Bremmer, author of *Collision Course*

"*The Genesis Machine* is fantastic, explaining how genetic code is the alphabet in which much of the future will be written. Amy Webb and Andrew Hessel have taken the very complicated subject of synthetic biology and made it understandable with sharp prose and sharp analysis that cut through mysteries of science and twenty-first-century humanism."

—Alec Ross, author of *The Industries of the Future* and *The Raging 2020s*

# THE GENESIS MACHINE

## ALSO BY AMY WEBB

*The Signals Are Talking:*
*Why Today's Fringe Is Tomorrow's Mainstream*

*The Big Nine: How the Tech Titans &*
*Their Thinking Machines Could Warp Humanity*

# THE GENESIS MACHINE

*Our Quest to Rewrite Life in the Age of Synthetic Biology*

AMY WEBB
AND
ANDREW HESSEL

PUBLICAFFAIRS
NEW YORK

PublicAffairs
Hachette Book Group
1290 Avenue of the Americas, New York, NY 10104
www.publicaffairsbooks.com
@Public_Affairs

Printed in the United States of America
First Edition: February 2022

Published by PublicAffairs, an imprint of Perseus Books, LLC, a subsidiary of Hachette Book Group, Inc. The PublicAffairs name and logo is a trademark of the Hachette Book Group.



Print book interior design by Linda Mark.

Library of Congress Cataloging-in-Publication Data
Names: Webb, Amy, 1974- author. | Hessel, Andrew, author.
Title: The genesis machine: our quest to rewrite life in the age of synthetic biology / Amy Webb and Andrew Hessel.
Description: First edition. | New York: PublicAffairs, 2022. | Includes bibliographical references and index.
Identifiers: LCCN 2021045281 | ISBN 9781541797918 (hardcover) | ISBN 9781541797932 (ebook)
Subjects: LCSH: Synthetic biology—Moral and ethical aspects. | Synthetic biology—Social aspects. | Genetics—Moral and ethical aspects. | Genetics—Social aspects.
Classification: LCC QH438.7 .W43 2022 | DDC 572.8/6—dc23/eng/20211012
LC record available at https://lccn.loc.gov/2021045281

ISBNs: 9781541797918 (hardcover); 9781541797932 (ebook)

LSC-C

Printing 1, 2021

*For Kaiya, sage and light. And for Steve, who rebooted me.*

—AW

*For Hani, Ro, and Dax, for the lessons on life.*

—AH

# *CONTENTS*

INTRODUCTION

# SHOULD LIFE BE A GAME OF CHANCE?

AMY—THE FIRST TIME I FELT THE SHARP TWINGE IN MY BELLY WAS during an important client meeting. Seated around the table were senior executives from a multinational information-technology company. We were developing the company's long-term strategy when the twinge hit me again. I quickly handed the meeting off to one of my colleagues and ran to the bathroom. By then, a layer of sticky, dark blood had soaked through my black tights and adhered to my inner thighs. I couldn't breathe. I couldn't physically take in air. I slumped over on the toilet and finally allowed myself to sob, silently, so no one could hear.

I'd been eight weeks along. I was scheduled the following week for an early ultrasound. I'd already started thinking of names: Zev if the baby was a boy, Sacha if she was a girl. As I cleaned up the blood on my legs and the floor, I searched for answers, but kept arriving at the same place of anger and self-blame. It was my fault. I must have done something wrong.

The third time I felt that twinge, I already knew what to expect: blood loss and a humiliating trip to the drugstore for extra-large pads, followed by deep depression, insomnia, and a stream of questions with no answers. My husband and I saw the best fertility specialists in Manhattan and Baltimore, and we subjected ourselves to every test offered: blood tests to evaluate my hormones, tests to make sure I had enough eggs in reserve, and tests to determine whether I had any benign growths or cysts that might be causing problems. These were high-tech guesstimates, not answers.

We kept trying, and in a subsequent pregnancy I made it past the four-month mark, a milestone, and we finally allowed ourselves to feel excited. We arrived at the OB-GYN for a routine checkup. I was at eighteen weeks now, and my belly was starting to protrude. I laid on the exam table, and a technician squirted cold jelly onto my midsection, smearing it around with a sonogram wand. She punched a few keys on her keyboard, zooming in on a grainy, mostly black video. She apologized, mumbled something about her old equipment, and left the exam room, returning with another machine and my doctor. Again, she squirted cold jelly onto me, and again smeared it around, clicking her keyboard to zoom in as she glanced at my doctor, and then, reluctantly, back at me.

I don't remember exactly what they said, but I remember my doctor taking my hand and the sound of my husband crying. I would be admitted for surgery to remove the fetal tissue. In the end, I was told that nothing was medically wrong with either of us. We were in our early thirties. We were healthy. We could get pregnant. The problem seemed to be my ability to stay that way.

One in six women will miscarry during her lifetime, and there isn't a singular reason. Most often, the cause is a chromosomal abnormality—something goes haywire as the embryo divides—that has nothing to do with the health or age of the parents. It wasn't my fault, I was told. My body just wasn't cooperating.[1]

**Andrew**—Since age ten, I had been resolute in my intention to never have children of my own. We'd lived on a rural farm property on the outskirts of Montreal. My parents struggled with each other, and as a result, with me and my two siblings. The three of us had been born in quick succession: my brother was a year younger than me, my sister a year older. When my parents told us they were separating, I wasn't upset, but I do remember thinking that my mom would have been happier as a nun. Instead, she became a single mother and a nurse working the night shift.

She slept during the day while we were at school. It helped that we were all independent, capable kids. I often escaped to the library—my second home, where I lived in the stacks. I'd bring home armfuls of books and see her off to work at 10 p.m., and I would keep watch over my siblings, often reading to them until dawn, when my mom returned home. Stories about traditional nuclear families felt foreign to me. I couldn't relate. What made sense was the dependable logic of engineering, the wonders of biology, and the visions of science fiction. Sometimes, when my brother and sister drifted off to sleep, I stayed awake reading and thinking about life: where creatures both enormous and microscopic came from, how they evolved, and the promise of what they might become.

By the time I was eighteen, I wanted to study the fundamentals of life—genetics, cell biology, microbiology—but I had no intention of making children of my own. At that point, I was writing software and databases, thinking in both genetic and computer code, and I had a lifetime of research ahead of me. Sex was compelling, kids were not. The only forms of male birth control were mechanical, not medical, and they were hardly reliable. The guaranteed solution was vasectomy, so I sought out my doctor and asked for one. At first, he protested—at age eighteen, I was barely an adult, and certainly in no position to make such a drastic choice. Vasectomies were reversible, I countered, and I could bank sperm if I had doubts, but I didn't. My conviction won his approval and referrals to urologists, but it would ultimately take six years to turn off the taps. Most of the specialists thought I was being rash and immature. I argued that I was just trying to be responsible. Still, once I got the vasectomy, there was no guarantee I'd be able to have children in the future.

Thirty years later, I connected with a beautiful woman at a conference who lit up when I talked about cells and who indulged my long-winded diatribes on DNA as software. Lying next to her one morning in her Manhattan apartment, I was overcome with a terrifying new feeling: I wanted children. I wanted that family, with her by my side. But I was now in my late forties, and I knew exactly what to expect medically and biologically.

When we decided to get pregnant, we were both hopeful, but realistic. On the day of my reversal, I fixed my eyes on the ceiling as the attendants pushed me into a surgical suite. The lights blurred past in a rhythmic pattern, and with each burst of light I cycled back to the doctor's warning so long ago, and thought about how life paths can suddenly change. The tubes connecting my testicles to my urethra, which would have enabled sperm to leave my body, hadn't been clamped or tied off—making a reversal easy. Instead, the surgeon had severed them entirely and cauterized them to make sure I didn't leak internally. It would take a delicate microsurgery and general anesthesia to reconnect them.

We tried, and failed, to get pregnant for eighteen months. I knew what was wrong—and how little I could do now to change things. The surgery had been successful, but my body's system had been shuttered for too long. Mechanically, there was nothing wrong with me. My body just wasn't cooperating.

RIGHT NOW, SCIENTISTS are rewriting the rules of our reality. The anguish we both experienced as we struggled to become parents could be an anomaly in the coming decades. An emerging field of science promises to reveal how life is created and how it can be re-created, for many varied purposes: to help us heal without prescription medications, grow meat without harvesting animals, and engineer our families when nature fails us. That field, which is called synthetic biology, has a singular goal: to gain access to cells in order to write new—and possibly better—biological code.

In the twentieth century, biologists focused on taking things apart (tissues, cells, proteins) to learn how they functioned. This century, a new breed of scientists is instead attempting to construct new materials out of life's building blocks, and many others are already achieving successes in the nascent field of synthetic biology. Engineers are designing new computer systems for biology, and startups are selling printers capable of turning computer code into living organisms. Network architects are using DNA as hard drives. Researchers are growing body-on-a-chip systems: picture a translucent domino embedded with nanoscale human organs that live and grow outside a human body. Together, biologists, engineers, computer scientists, and many others have forged a genesis machine: a complex apparatus of people, research labs, computer systems, government agencies, and businesses that are creating new interpretations, as well as new forms, of life.

The genesis machine will power humanity's great transformation—which is already underway. Soon, life will no longer be a game of chance, but the result of design, selection, and choice. The genesis machine will determine how we conceive and how we define family, how we identify disease and treat aging, where we make our homes, and how we nourish ourselves. It will play a critical role in managing our climate emergency, and eventually, our long-term survival as a species.

THE GENESIS MACHINE incorporates many different biotechnologies, all of which were created to edit and redesign life. A series of new biological technologies and techniques, which broadly fall under synthetic biology's umbrella, will allow us not just to read and edit DNA code but to *write* it. Which means that soon, we will program living, biological structures as though they were tiny computers.

It's been possible to edit DNA code since the early 2010s using one of those technologies: CRISPR-Cas9.[2] Scientists refer to a pair of "molecular scissors" to describe the technique, because it uses biological processes to cut and paste genetic information. CRISPR routinely makes headlines about groundbreaking medical interventions, such as editing the genes of blind people to help them see again. Scientists have been using CRISPR's physical molecular scissor technique and splicing the DNA molecule back together in a sort of biological collage with letters rearranged into new places. The problem is that researchers can't directly see the changes being made to the molecule they're working on. Each move requires laboratory manipulations that must then be experimentally validated, making it all very indirect, labor intensive, and time consuming.

Synthetic biology digitizes the manipulation process. DNA sequences are loaded into software tools—imagine a text editor for DNA code—making edits as simple as using a word processor. After the DNA is written or edited to the researcher's satisfaction, a new DNA molecule is printed from scratch using something akin to a 3D printer. The technology for DNA synthesis (transforming digital genetic code to molecular DNA) has been improving exponentially. Today's technologies routinely print out DNA chains several thousand base pairs long that can be assembled to create new metabolic pathways for a cell, or even a cell's complete genome. We can now program biological systems like we program computers.

These scientific innovations have fueled the recent and rapid growth of a synthetic biology industry intent on making high-value applications that include biomaterials, fuels and specialty chemicals, drugs, vaccines, and even engineered cells that function as microscale robotic machines. Progress in artificial intelligence has provided a significant boost to the field, as the better AI becomes, the more biological applications can be tested and realized. As software design tools become more powerful and DNA print and assembly technologies advance, developers will be able to work on more and more complex biological creations. One important example: we will soon be able to write any virus genome from scratch. That may seem like a frightening prospect, given that the coronavirus known as SARS-CoV-2, which causes COVID-19, has, as of this writing, resulted in the deaths of more than 4.2 million people worldwide.[3]

What makes viruses like SARS-CoV-2—and SARS, H1N1, Ebola, and HIV before it—so difficult to contain is that they are powerful microscopic code that thrive and reproduce with an unprotected host. You can think of a virus as a USB stick you'd load into your computer. A virus acts like a USB by attaching itself to a cell and loading new code. And as bizarre as this might sound at a time when we're living through a global pandemic, viruses could also be our hope for a better future.

Imagine a synthetic biology app store where you could download and add new capabilities into any cell, microbe, plant, or animal. Researchers in the United Kingdom synthesized and programmed the first *Escherichia coli* genome from the ground up in 2019.[4] Next, the gigabase-scale genomes of multicellular organisms—plants, animals, and our own genome—will be synthesized. We will someday have a technological foundation to cure any genetic disease in humankind, and in the process we will spark a Cambrian explosion of engineered plants and animals for uses that are hard to conceive of today, but will meet the global challenges we face in feeding, clothing, housing, and caring for billions of humans.

Life is becoming programmable, and synthetic biology makes a bold promise to improve human existence. Our purpose in this book is to help you think through the challenges and opportunities on the horizon. Within the next decade, we will need to make important decisions: whether to program novel viruses to fight diseases, what genetic privacy will look like, who will "own" living organisms, how companies should earn revenue from engineered cells, and how to contain a synthetic organism in a lab. What choices would you make if you could reprogram your body? Would you agonize over whether—or how—to edit your future children? Would you consent to eating GMOs (genetically modified organisms) if it reduced climate change? We've become adept at using natural resources and chemical processes to support our species. Now we have a chance to write new code based on the same architecture as all life on our planet. The promise of synthetic biology is a future built by the most powerful, sustainable manufacturing platform humanity has ever had. We're on the cusp of a breathtaking new industrial evolution.

The conversations we're having today about artificial intelligence—misplaced fear and optimism, irrational excitement about market potential, statements of willful ignorance from our elected officials—will mirror the conversations we will soon be having about synthetic biology, a field that is receiving increased investment because of the novel coronavirus. As a result, breakthroughs in mRNA vaccines, home diagnostic testing, and antiviral drug development are accelerating. Now is the time to advance the conversation to the level of public consciousness. We simply do not have the luxury of time to wait any longer.

The promise of this book is simple and straightforward: if we can develop our thinking and strategy on synthetic biology today, we will be closer to solutions for the immediate and long-term existential challenges posed by climate change, global food insecurity, and human longevity. We can prepare ourselves now to fight the next viral outbreak with a virus we engineer and send into battle. If we wait to take action, the future of synthetic biology could be determined by fights over intellectual property and national security, and by protracted lawsuits and trade wars. We need to ensure that advances in genetics will help humanity, not irrevocably harm it.

The code for our futures is being written today. Recognizing that code, and deciphering its meaning, is where humanity's new origin story begins.

This is a book about life: how it originates, how it is encoded, and the tools that will soon allow us to control our genetic destinies. It is also about the right to make decisions about life, defined for a new generation along scientific—as well as ethical, moral, and religious—terms. With powerful systems in place, to whom will we grant the authority to program life, to create new life forms, and even to bring former life forms back from extinction? Answering these questions will force humanity to resolve economic, geopolitical, and social tensions.

- Those who can manipulate life can exert control over our food supply, medicines, and the raw materials required for our survival.
- Our future health and prosperity will be determined, at least in part, by the companies that invest in and control the legal rights to genetic code and to the processes by which it is altered.
- Genome editing and DNA synthesis are the cornerstone technologies of synthetic biology, and the global market for these tools is booming. However, there are looming disagreements about whether these tools, and our raw genetic data, should be accessible to everyone, or should instead be held in proprietary databases, and licensed to those who can afford access.
- Venture-backed startups cannot return investments from basic research alone, so there is often pressure to develop marketable products within reasonable time frames. While privately funded companies have freedom to innovate, publicly funded biotechnology research tends to move slowly, adhering to traditional practices.
- Absent a mandate, such as winning the space race or deploying an effective vaccine, government grants reward competencies and conservatism; they do not incentivize speed, innovation, or forward-leaning approaches.
- Those who legislate, write policy, create and enforce regulations, and enact laws wield tremendous power over our futures, and at present there is no consensus on the acceptable circumstances under which humans should manipulate human, animal, or plant life.
- Nor is there consensus on how to make decisions that could benefit us on a planetary scale. In the United States, entirely new life forms

that have never existed before are already in development—some have been booted up from computer code to living tissue.

- In China, President Xi Jinping has proclaimed that China "must vigorously develop science and technology and strive to become the world's major scientific center and innovative highland," with a primary focus on rewriting life.[5] China's strategic roadmap includes a comprehensive database for genomic information and an aggressive time frame for the commercialization of engineered living systems. The country's leadership seeks to move up the value chain from "workshop of the world" to become the global leader in modern industries such as both biotechnology and artificial intelligence.[6]
- The United States and China may be interdependent and reliant on each other's economies to prosper, but China's quest to become the prevailing technological, scientific, and economic superpower has long caused tension between the two countries. A coordinated, enforceable plan is table stakes, because our current geopolitical tensions do not mirror past conflicts.
- The ability to edit and write life has profound societal impacts, and we must balance the public's trust and the speed of biotechnological progress. We will need to reconcile our desires for privacy with the advancements brought by enormous datasets made of our genetic code.
- We must determine how to make this technology equitable and accessible to all, but a division is inevitable, because not everyone will trust science or have access to the very latest tools. For that reason, we will need to prepare for difficult societal issues, such as how to manage a genetic divide. Part of this divide will be between people with enhanced genetic codes—who may have special abilities, or upon whom special privileges may be bestowed—and people who have never had their codes manipulated.

This book is also about you and your life, and the decisions you will need to make in your lifetime. We are standing on the precipice of sweeping change, and you must play an active role in your own future by making informed decisions today. You will need to make choices that have consequences, such as whether to have your own genome sequenced and what to do with that data. Or, if you are planning to become a parent, whether to freeze your eggs, pursue assistive reproductive technology such as in vitro fertilization (IVF), or use genetic screening to select the strongest of your embryos. These are decisions with which we are intimately familiar. In fact, they are what compelled us to write this book.

In order to see what future the genesis machine might someday build, it's important to revisit the past. In the first part of this book, we'll explain synthetic biology's origin and the history of how researchers decoded life—and eventually manipulated it—with the intent to create synthetic organisms whose parents were computers. In Part Two, we will reveal the new bioeconomy created by the genesis machine, which includes the myriad, fantastical medicines, foods, coatings, fabrics, and even beers and wines that entrepreneurs are attempting to make—as well as the possible biotechnology solutions to solve problems such as the spread of ocean plastics, the increase in extreme weather events, and the ongoing possibility of dangerous viruses that could lead to new pandemics. We'll also address the risks that synthetic biology poses, which range from cyberbiology hacking to the looming genetic divide, pitting wealthy engineered people against those who will not be able to afford technology-assisted reproduction. In Part Three we'll explore different futures in the form of creative, speculative scenarios suggesting the many ways in which the genesis machine might transform the world. Finally, in Part Four we offer our recommendations for ensuring that the genesis machine gives birth to the best of these possible futures.

But first, you should meet a young man named Bill.

# PART ONE | Origin

| ONE |

# SAYING NO TO BAD GENES

## *The Birth of the Genesis Machine*

THE LONG DAYS HAD SHORTENED AND COOLER NIGHTS HINTED AT autumn in Duxbury, Massachusetts, a pretty seaside town just south of Boston. Bill McBain was a gifted student with wide-ranging interests in photography, math, and journalism, but in other ways he was unremarkable: on the first day of eighth grade, it was obvious that Bill had gone through a growth spurt over the summer, just like his friends. He was now four inches taller. But unlike the other kids, he'd also lost weight. While his male friends were starting to fill out and put on some adolescent muscle, Bill was spindly—all elbows, ribs, and knees.

Bill went to bed early every night and woke up exhausted every morning. He started drinking water—lots of it—but couldn't seem to ever quench his thirst. It was 1999, and transparent plastic Nalgene water bottles—intended for rugged outdoor use—were suddenly wildly popular as a fashion accessory at school. But for Bill, his Nalgene was a necessity: he filled it with water between classes and guzzled from it continuously. Once, while staring at the ounces marked on one side of the bottle,

Bill—who loved math—drifted off, making some mental calculations. He estimated that he was drinking four gallons of water a day, sometimes five.

In February, a family friend visiting for the afternoon watched nervously as Bill gulped from his water bottle again and again. As a nurse, she immediately recognized the warning signs and made a quick, discreet trip to the bathroom to confirm her hunch: indeed, the toilet seat was sticky to the touch, and when she bent over to take a whiff it was sickly sweet. She asked Bill's parents to take him to the clinic to get his blood tested the next morning.

On the way there, the family pulled over for a quick breakfast, and Bill ordered a cinnamon sugar bagel along with a large red Gatorade to wash it down. It wasn't the best meal to eat before a fasting blood sugar test, but he didn't know any better. At the clinic, the doctor pricked Bill's finger with a tiny needle and squeezed a droplet of blood onto a test strip attached to a meter. Within a few seconds the meter beeped, and the screen flashed "high." This meant that his blood sugar level had spiked above 500 milligrams per deciliter (mg/dL). The fasting blood sugar of someone with a normal pancreas would typically fall between 70 and 99 mg/dL, or just below one-thousandth of a gram per one-tenth of a liter. In other words, barely noticeable, because a healthy person's system quickly breaks down sugar and converts it into energy, so there isn't much left in the bloodstream. If a healthy person takes that same blood test right after eating, the number will be higher for a few hours, as their body processes the food, but it will still be less than 140 mg/dL.

The doctor drew more blood and sent it to his lab for a detailed analysis. The results left him at a loss for words. Back in his office with Bill and his parents, he sat down. He looked from his file folder to Bill and his parents, and then back to his folder again. Bill's blood sugar reading was a staggering 1,380 mg/dL. His sodium, magnesium, and zinc levels were so far off the charts that the pH of his blood had actually changed. He was on the verge of sinking into a diabetic coma, if not worse: blood like that could kill.

Bill and his parents were forced into a crash course about the mechanics of type 1 diabetes and how to treat it. A healthy pancreas is always slowly secreting insulin, the hormone required of our cells to make energy. When you eat, your pancreas gives you an additional, bigger boost of insulin to metabolize the sugar you've consumed. But Bill's pancreas had suddenly stopped producing insulin. Type 1 diabetes typically shows up during adolescence, and he had all the classic symptoms: fatigue, excessive thirst, sticky-sweet urine and continually needing to go to the bathroom. The urge to continuously drink water was his body's crude attempt at self-treatment: lots of water would help flush out the unmetabolized sugar in his blood. Eventually, though, he would face a life-threatening chain reaction. His body would start using fat for the energy it needed to stay alive, and in the process, chemicals called ketones would be released. Ketones, which are highly acidic, would get stuck in Bill's bloodstream, where they are a poison. When the levels got too high, Bill would wind up in diabetic ketoacidosis—otherwise known as diabetic coma. At that point, if he was left untreated, death would come quickly.

Worried that they'd somehow contributed to his diagnosis, Bill's parents asked what had caused Bill's condition. The hurried bagel and Gatorade breakfast wasn't typical, they assured the doctor; the family usually ate healthy meals and got lots of exercise. "It's just bad genes," the doctor told them. Scientists didn't know exactly why for some people the body became resistant to insulin, he said, or why, in some adolescents—like Bill—the pancreas suddenly stopped functioning properly. But there was a silver lining: a treatment regimen to manually perform all the tasks that his body should have been doing automatically. Bill would need to start injecting himself with a drug called Humulin Regular, a synthetic human insulin that could deliver short bursts around mealtimes, and Humulin NPH (*neutral protamine hagedorn*), which was designed to give him a slow drip of insulin overnight.[1]

## DISCOVERING INSULIN

The clinical symptoms related to type 1 diabetes—frequent urination, confusion, irritability, difficulty concentrating, and sometimes death—were first recorded some 3,000 years ago in Egypt. In around 1550 BCE* it was an Egyptian who recommended drinking "a measuring glass filled with

* BCE stands for "before Common Era" and is an alternative to BC, which stands for "before Christ."

Water from the Bird pond, Elderberry, Fibres of the asit plant, Fresh Milk, Beer-Swill, Flower of the Cucumber, and Green Dates" as a treatment for excessive urination. The Egyptian physicians already suspected that there was a connection between what people ate and the symptoms we now associate with diabetes. But it was another 1,500 years before Aretaeus, a Cappadocian physician who spoke Greek, described "a melting down of the flesh and limbs into urine," a condition he named *diabetes* after the Greek word for "siphon." At about the same time, physicians in China and South Asia were making similar discoveries.[2]

In 1674, a physician at Oxford University named Thomas Willis began pursuing his own research using a procedure which will sound pretty gross. He had patients with symptoms of diabetes urinate into a little glass, from which—you might want to skip the rest of this paragraph if you're eating—Willis would sniff and sip. Like the electronic monitor that evaluated the milligrams of sugar per deciliter in Bill's blood, Willis was checking for elevated sweetness.[3]

But a clear understanding of what caused diabetes remained elusive for the next few centuries. Some doctors in the early 1900s advocated what they called a "starvation diet," theorizing that if patients were denied all forms of sugar, the diabetes might go away on its own. Unsurprisingly, this led to worse problems—patients tended to starve to death rather than improve.

Then, in 1921, there was a breakthrough.[4] By that point, a long-standing—if unproven—theory in the medical community held that a secretion from the pancreas was responsible for regulating blood sugar, and Canadian physician Frederick Banting and his student Charles Best had begun hypothesizing that digestive enzymes might be destroying that secretion before any researcher could extract it. They planned to tie off the pancreatic ducts until the cells that produce the enzymes degenerated, then analyze what was left over.[5] Unfortunately neither of the men were trained as surgeons and their early research, conducted on laboratory dogs, was frankly grisly: most of the dogs died. So they started buying stray dogs off the black market instead, and with some practice, they succeeded in removing a pancreas without killing the animal. They then froze it, ground it into a paste, filtered it, and injected that liquid back into the dog. They took blood samples every thirty minutes to observe whether its blood sugar changed at all. To their astonishment, the dog's blood sugar returned to normal levels—even though the poor pooch now had no pancreas. They observed measurable changes in what would later become known as insulin.[6]

If the treatment worked in dogs, might it also work in humans? It might. But finding a healthy human cadaver's pancreas—to say nothing of finding thousands of them on an ongoing basis, to feed new demand should the treatment work—presented obvious problems. So Banting and Best, along with a newly augmented research team, turned to cattle. They ordered pancreases from a local meatpacking house and ran them through an industrial grinder: picture an enormous machine helmed by someone wearing oversized gloves who's shoving gland after gland into a funnel on top, and a spout at the bottom extruding pulverized tissue into a bin.

They extracted insulin, purified it, and injected it into an adolescent boy like Bill—he was fourteen years old, with juvenile diabetes, and would have died without an intervention. The teenager improved dramatically. In a stroke of magnanimity and foresight, the research team offered licenses to pharmaceutical companies permitting them to reproduce their work for free. It galvanized the commercial production of insulin. Banting, Best, and their research team won the Nobel Prize in 1923, in recognition of how their work had changed the course of life for millions of people worldwide.[7] But over the years, the ranks of diabetics kept growing, and there were only so many cow pancreases that could be harvested.

## THE BIRTH OF BIOTECH

Injecting bovine insulin addressed—but didn't actually solve—the "bad gene" problem that Bill's doctor had referenced. Nor did it help the growing number of adults diagnosed with type 2 diabetes. In people who develop type 2, researchers blame environmental factors, including obesity, inactivity, and indulging in too many sweets, as well as a predisposition to the disease. That's why seemingly fit, athletic people can also mysteriously develop the same warning signs Bill did. There are theories describing what might be going wrong: sometimes the body's own immune system, which normally fights harmful viruses and bacteria, gets confused, and mistakenly starts destroying insulin-producing cells instead. Other theories blame a diabetes-causing virus, or suggest that it could be a second-order effect of a virus that's silently attacking the body in other ways. The standard treatment for the past one hundred years has been to ask patients to track exactly what they are eating and how much energy they are expending, either through brute force counting or, more recently, with the help of a digital glucose monitor. Some form of medication, including insulin and pills, adjusts the level of blood sugar to a normal range.

How did we evolve from smashing and extruding a cow pancreas for its insulin to the high-tech pump and synthetic human insulin Bill uses today as an adult? Shortly after Banting and Best proved that bovine insulin worked, pharmaceutical company Eli Lilly began manufacturing it, but in 1923, the process was slow, expensive, and resulted in an unforeseen supply chain problem: names being added to waiting lists for insulin drastically outpaced farmers' ability to raise and slaughter their cattle herds.[8] Researchers found other options that worked in humans—extruding pig pancreases resulted in usable insulin—but there was no sustainable way to manufacture supplies at a reasonable scale. It took 8,000 pounds of pancreas glands—which required harvesting pancreases from roughly 23,500 animals—to make just one pound of insulin. That amounted to about 400,000 vials of insulin, which was only enough to treat 100,000 patients per month. But that wasn't much, given rising demand.[9] Roughly 1.6 million people needed insulin by 1958; the number topped 5 million in the United States alone by 1978.[10] That meant Eli Lilly would need to harvest pancreases from 56 million animals a year just to provide enough insulin for Americans. The company had to find an alternative, and fast.

Just before he died in 1977, Eli Lilly Jr., whose grandfather had founded the company that bore his name, launched a strategic initiative to solve the pancreas problem.[11] If cows and pigs could be used, surely there were many other viable animal candidates. He cut deals with several universities, including Harvard and the University of California at San Francisco, to develop new insulin prototypes from other animals. These institutions began working on rat versions of the insulin gene. Lilly

Jr. promised a lucrative contract to the first institution that could solve this supply problem and finally speed up insulin production.[12]

But another group of researchers had a radically different idea for the future, one that didn't involve organ harvesting at all. If there was no cure for diabetes, and if the number of people diagnosed with it continued to increase, Eli Lilly—not to mention all the other pharmaceutical giants—would at some point face another supply chain issue. From this group's perspective, there were actually two addressable problems spanning a longer time horizon. The first problem—the supply issue—could be solved by having engineered bacterial cells produce human insulin, rather than growing and squeezing it out of livestock. The second problem, which could be addressed in the future, was reprogramming "bad genes" to behave properly. Harvard, UCSF, and a startup called Genentech were all using rDNA technology. The thing that separated Genentech from the other groups was that they decided to go straight to cloning and expressing human insulin in E. coli.

Genentech had only been in business for a year, and the researchers there were working on a controversial new technology called recombinant DNA. While the established universities and pharmaceutical companies, teeming with decorated biomedical scientists, were refining well-worn practices, Genentech was instead tinkering at the molecular level, taking two different strands of DNA and "re" combining them together.[13] DNA, or deoxyribonucleic acid, is the genetic material of life, and recombinant DNA technology makes it possible to splice divergent species—for example, human and bacterial—together to replicate, synthesize, and potentially improve our existing genetic code.[14]

While Genentech had already scored some early successes by 1977, it wasn't taken seriously by the research establishment. There were a few reasons. First, "synthesizing" was akin to "cloning" genetic material, and that could lead to downstream risks, such as genetic manipulation. Given the progress being made in another controversial technology—in vitro fertilization, or IVF—some foresaw a future in which humans created designer babies with desired hair and eye color, musculature, and other traits. At that point, there was wild, dystopian speculation and a staunch resistance to change.[15] As a result, Genentech's recombinant DNA technology was deemed highly unorthodox and in need of added scrutiny.

To make matters worse, Genentech's biotech research funding came from venture capitalists rather than the federal government, another red flag to the establishment. The startup venture capital firm then known as Kleiner Perkins Caufield & Byers invested a reported $1 million in seed funding in Genentech (roughly $4.6 million today adjusted for inflation).[16,17] The partners were new to their field, too, and had been primarily interested in semiconductors. They took a chance on Genentech's vision for the future—and Genentech took on the risk of working with funders who, unlike the federal government, would require a return on their investment.

As a startup, Genentech spent no money on creature comforts. Around the same time Steve Jobs and Steve Wozniak were building computers in a garage, Genentech's team of scientists built a biochemistry lab in an air freight warehouse in a resolutely unpretty stretch of industrial South San Francisco. Genentech scored some early successes using their recombinant DNA techniques. The company's labs synthesized another pancreatic hormone—somatostatin, which helps regulate the endocrine system. When word spread that Eli Lilly had launched its insulin challenge, Genentech thought it might have a viable—if wildly different—solution to the supply problem.

Given that Genentech's recombinant DNA approach challenged conventional thinking, there weren't exactly dozens of research universities offering to partner with it or dedicate their labs to the work. If Genentech was going to compete, it would need to recruit more scientists willing to push the boundaries of recombinant DNA for insulin production. The potential rewards were enormous, but this wasn't a contest with silver and bronze medals: Eli Lilly was only interested in which team delivered a safe, scalable product. Genentech would either come in first, and win the contract—or it wouldn't, and end up empty-handed after all its feverish work.

The experiment would require round-the-clock work to advance the gene-splicing technique that Genentech had first developed in making its somatostatin discovery. Lilly provided added funds, and the founders expanded its team with young scientists fresh out of graduate school. Rather than the usual cohort of researchers, Genentech put together a supergroup with a wide range of specialties, including organic chemists

(Dennis Kleid and David Goeddel, who had been working on cloning DNA at Stanford Research Institute), a biochemist (Roberto Crea, who specialized in modifying nucleotides), a geneticist (Arthur Riggs, who expressed the first artificial gene in bacteria), and a molecular and cellular biologist (Keiichi Itakura, who had helped develop recombinant DNA technology).[18,19]

The challenge Genentech faced in synthesizing the insulin molecule was that it had long strings of amino acids—fifty-one, instead of somatostatin's fourteen—and there were two chains, A and B, that were chemically linked together. They'd have to assemble the correct bits of DNA code to make each chain, transplant them into two different bacterial strains, and hijack the cellular machinery of the bacteria to synthesize the chains. And this would only get them halfway there. To the Genentech team, proteins—which catalyze most of the reactions in living cells, and control virtually all cellular processes—were the key to cultivating insulin.

But assuming the team managed to get fifty-one amino acids—the molecules that combine to produce protein—in precisely the right order, in order to manufacture insulin, they'd still need to re-create them.[20] This would require chemically linking together the correct snips of DNA sequences, stitching them together, transplanting them into bacteria, and forcing the machinery of bacteria to produce their synthesized insulin chains—no easy task. If all went well, they would still need to purify the insulin chains, recombine them to form a complete molecule, and then hope that it was identical to the molecule produced by a human pancreas.

It was a moon shot on a cellular level, being attempted by a tiny group of under-resourced research scientists whose ideas about the future were mystifying to some, and downright dangerous to others. The complexity of the task, and the scope of their competition, forced the Genentech team to work in secret from their homes, using borrowed time in labs, and in a forgotten warehouse, far from the hallowed halls of Harvard and the University of California, all while coping with immense stress and an unforgiving deadline. First, the team would need to build a synthetic gene with exactly the right sequence of DNA that could act as instructions for a protein. Then, they'd have to transfer that gene to the right place of an organism that could read the instructions and produce the desired protein, which in this case was insulin.

The team painstakingly mixed chemicals and tested different permutations, again and again, trying to build the right sequence of DNA strands. They also needed to work with the bacteria itself, to figure out where exactly to splice the *E. coli* with the synthetic gene to produce the protein they needed. The process was analogous to a baking show challenge. Imagine the judges giving you a box full of ingredients, another box full of kitchen utensils, and an oven and telling you to bake a twelve-layer chocolate cake—under a crazy tight reality show deadline, working in an old run-down kitchen, and without any instructions.

But in the early morning hours of August 21, 1978—well ahead of their peers, and to the great surprise of everyone (including their own team members)—they pulled a perfect cake out of the oven.[21] Genentech had managed to produce the exact DNA sequence, instruct an organism to execute commands, and produce human insulin. It was the birth of biotechnology and the genesis of a new field of science called *synthetic biology*.

Lilly signed a multimillion-dollar, twenty-year contract with Genentech to develop and market the world's first biotechnology product: Humulin, which won US Food and Drug Administration (FDA) approval in 1982.[22]

## THE FACTORY OF LIFE

Genentech's truly astonishing achievement set human society on a different course. For the first time, we intervened in a biological process to manipulate cells and molecules, to overwrite what the body would do naturally. In healthy people, our cells are analogous to a futuristic, automated, computerized factory operating at the highest levels of efficiency. Imagine networks of advanced robots all working together: 3D printers that manufacture everything needed on demand at any quantity; a supply chain and logistics system optimized for maximum output; and an operating system with billions of lines of code that are all executed continuously. In the history of human society, no machine or factory we've ever built is as advanced or as elegant. Your body is simply a mobile giga complex—

housing for nearly forty trillion futuristic cellular factories all working together to keep you alive.[23]

Each one of those cellular factories has three main components: a set of instructions, a communications system to transmit those instructions, and a production line that makes the designated product. Those components are DNA, RNA, and protein. The inconceivably vast genetic ecosystem responsible for all life forms is only made up of these three primary molecular actors.

In biology class, we all learn about the twisted ladder of DNA's double helix. It's unmistakable and iconic, and it's made up of nucleotides represented by the letters A (adenine), T (thymine), G (guanine), and C (cytosine) chemically linked to a sugar (deoxyribose) and phosphate (acidic) backbone. These nucleotides, when paired, snap together tightly. But they can also be tugged apart relatively easily. This allows the two sides of the DNA double helix to split apart, much like how a zipper unzips. When DNA unzips, a cell can make precise copies of its DNA by using the unzipped DNA as a template to write new complementary strands before zipping up both of the strands again. The order, or sequence, of the four nucleotides in the DNA chain encode all the information the cell needs to live and thrive. DNA stores our genetic instructions, and while other microbes (such as viruses) are capable of carrying their own set of instructions, within cells it's the DNA that runs the show. It's not a stretch to say that the DNA molecule is likely the most significant molecule of all time (although water and caffeine undoubtedly have their partisans).

The DNA stores genetic instructions in cells, but it requires ribonucleic acid, or RNA, to tell the cell's factory what the DNA wants it to do. The RNA is converted, or translated, into a sequence of amino acids in a complex machine within a cell called a ribosome. As the RNA makes its way into the ribosome, a magical process takes place. Messenger RNA, or mRNA, attaches itself to the ribosome and looks for the biological equivalent of the "start" button, a three-letter sequence called a codon. The ribosome works through the strand of mRNA, reading each three-letter set, until it finds the "stop" button. All the while, the cellular factory's product—protein—is being produced.

Proteins—those chains of amino acids—are the main structural material of cells and do most of the operational work, and there are thousands of different types offering a range of functions. Structural proteins, such as collagen, provide for tendons and cartilage, for example. Hemoglobin is a transport protein, which carries essential oxygen in red blood cells. Antibodies are Y-shaped proteins with special recognition capabilities: when they encounter a microbe for the first time, they attach to it and either collaborate to destroy it or block it from infecting other cells. If you've recovered from an infection, a small number of antibody-producing immune cells stay in your body as memory cells, and they jolt back into action the next time you encounter the same microbe that caused the infection. Vaccines are designed to trigger the same response. Although there are more than five hundred known amino acids, just twenty routinely show up in biological systems.[24]

If the cell is a futuristic factory, the genome is a futuristic operating system where genes can be switched on or off. Two organisms might both have the same gene linked to a particular trait, but if that gene isn't switched on, it won't be expressed. The control of which genes are turned on and off, or how much, is complicated and regulated. It involves non-protein coding sequences such as promoters and enhancers and various protein transcription factors. It has been difficult to study them because these factors are hard to measure in real time, but here's a wild example: the winter skate—a flat, cartilaginous species of fish—automatically switches on its genes to change its body structure to adapt to increasingly warmer winter waters brought by climate change.[25]

Unlike a traditional factory or a traditional computer, where the logic and structural machinery are independent, the operating system for life requires full interoperability—and we're just starting to interpret how it all works together. For example, a new PC would have the latest version of Windows installed, but you would need to purchase games and productivity software separately and load it into the machine. That's not the case in biology, where the machine and the information are fully intertwined.

Today's electronic computers are still little more than fancy calculators. They also gobble energy, are brittle, can't repair or manufacture themselves, and can't produce anything tangible without connecting to a printer. Cells are the computers that computers, if they could dream, would dream of becoming: computers that can self-manufacture, self-repair, and run on almost any energy source.

That is precisely why Genentech's pioneering work was so profound, and why synthetic biology will reshape life as we know it today. Once we can speak and manipulate the language of biology, we will have a say in what's happening inside of cells. We'll have the power not just to read the code and edit it—either to clone insulin or make minor repairs—but to write new instructions, and to have those instructions delivered, and produce new biological products on the other side. Humulin was an early product of synthetic biology, a field that is still very new but is growing. Researchers working in the field struggle to define its contours, but it is an umbrella for chemistry, biology, computer science, engineering, and design united for a single goal: to gain access to the cellular factory and to life's operating system in order to write new—and possibly better—biological code.

Synthetic biology intersects with computer science and, notably, artificial intelligence, making use of machine learning and uncovering meaningful patterns in big sets of data. Machine learning powers the services you use all the time, like the recommendations you get on YouTube and Spotify, and the interactions you have with voice assistants like Alexa and Siri. In a biology context, machine learning empowers research scientists to pursue myriad small bets on patterns. Running experiments with several variables often requires tiny, methodical tweaks to measurements, materials, and inputs—and at the end of the process, there still might not be a viable product. Google's DeepMind division, which researches and builds AI systems that are then unleashed on knotty problems, developed a way of testing and modeling the complex folding patterns of long chains of amino acids, thus solving a problem that has long vexed scientists. The DeepMind system that accomplished that—AlphaFold—was used to predict the structure of more than 350,000 proteins from humans and 20 model organisms. The dataset is expected to surpass 130 million structures by 2022.[26] This will allow scientists to develop drugs to treat diseases far more quickly than the trial-and-error method Genentech used to create Humulin.[27] This technique, and other synthetic biology approaches, can lead to labs making better bets and more of them, which lowers the cost of bringing new drugs to market.

The Genentech researchers synthesized human insulin *before* the era of artificial intelligence and computers using massive datasets, machine learning, and deep neural nets developed to outthink humanity's smartest people. Today there are vast databases of proteins and metabolism and computers capable of running billions of simulations, over and over, in search of a solution to computational problems. If the same group of researchers set out to solve the insulin problem now, they wouldn't need months of working around the clock in a lab, hunched over test tubes and petri dishes. Working alongside an AI-powered platform, they could run all the different possible combinations of those three-letter codes to design the ideal solution within a matter of hours.

Forty trillion microscale factories follow instructions, make decisions, replicate, and communicate with each other autonomously throughout your day without ever asking for your permission or input. Within the next decade, synthetic biology will put the power to program the ultimate supercomputer—cells—in human hands.

## OVERWRITING BAD GENES

What if we challenged a deeply held assumption: that bad genes, like the ones that cause Bill's type 1 diabetes, are simply an unfortunate fact of the human condition? Bill was lucky. His parents knew how to get him very good care, and—more importantly—they could afford it. His condition became a family project. After the school year ended, his parents enrolled him in diabetes camps in the summer, where he spent time around other kids and doctors learning how to manage his diabetes. But even today, someone like Bill, who went to a special camp and had parents who were vigilant about his health, would still face uncertainty about his diabetes.

At the height of the COVID-19 pandemic, millions of Americans lost their jobs and their health coverage. New underground sharing networks for people with diabetes popped up on Facebook: insured people with extra vials of insulin began supplying others who could not afford it and would otherwise die.[28,29] These weren't Silk Road–style, back-alleys-of-the-internet drug deals. These people were working the system to help save lives. But even before COVID-19, 25 percent of people with diabetes in the United States were forced to ration their insulin because of the price.[30] (This is especially true of Latino, Indigenous, and Black populations, groups that have elevated rates of both diabetes and poverty.) Before the pandemic triggered border closings, American diabetics often crossed to Mexico or Canada to buy insulin at a fraction of its cost in the US.[31]

Insulin, which an estimated 10 percent of Americans require every day, is only made by three companies—Sanofi, Novo Nordisk, and Eli Lilly—and its price has skyrocketed.[32,33] Between 2012 and 2016, the cost per month doubled, from $234 to $450.[34] Today, one vial of insulin can cost $250. Some people need six vials per month, which sometimes forces Americans who lack good health insurance to ration their doses—or to choose among paying for insulin, feeding their families, or paying rent. Pharmaceutical companies argue that rising prices reflect the cost of innovation. Creating ever more effective formulas, tests, and technologies costs money and time—as we saw with Genentech and Banting and Best—and as publicly traded companies they must earn back the investment they make in R&D.

There's a historical irony here. Recall that when Banting, Best, and their team first discovered and created insulin back in 1923, they refused to commercialize or earn money off their discovery. They sold the patent to the University of Toronto for $1, because they wanted everyone who needed the life-saving medication to be able to afford it. "As solutions to the insulin-cost crisis are being considered," wrote the *New England Journal of Medicine* editorial board, "there is value in remembering that . . . [both Banting and Best] felt that insulin belonged to the public. Now, nearly 100 years later, insulin is inaccessible to thousands of Americans because of its high cost."[35]

Today's insulin is manufactured in factories using a synthetic process, but it was designed to simply mimic what the body must do on its own. As synthetic biology evolves, we won't be limited to imitation: a customized strain of insulin-producing cells could be engineered to work in far more precise and sophisticated ways. One of the most promising developments centers around re-engineering cells that can manufacture insulin only when needed. The implications are profound: What if, in the future, vials of expensive insulin were no longer necessary? What if, instead of insulin pumps and injections, diabetics instead took a one-time dose of synthetic cells that could react to blood glucose levels, and then produce insulin on their own?

Although it sounds like science fiction, that future is closer than you might think. In 2010, one of the world's preeminent biotechnologists, John Craig Venter, led a team that synthesized the DNA of a whole bacterium—copying something that already existed in nature—but they added a twist. The new genome included the names of the forty-six researchers who helped write the project, along with quotes from J. Robert Oppenheimer, poems from James Joyce, and secret messages that could be decoded, like a puzzle. When the bacteria reproduced, it carried this new biological code—and the poems, quotes, and messages—from generation to generation. It was the first evidence that a new form of life, programmed to complete designated tasks, could be created, and could thrive.[36]

This wasn't merely synthesizing human insulin. It was a designed, intentional evolution of life developed using a computer-generated genome. We saw a glimpse of that power in 2019, when researchers working with Venter demonstrated that it was possible to write genetic code, portending a future in which it would be possible to improve on the genetic hand that people like Bill were dealt.[37] In other words: if cells can be reprogrammed, then maybe people with diabetes could become their own internal pharmacies.

The broader implications are both profound and problematic: If a group of scientists can create a new strain of bacteria biologically watermarked with "To live, to err, to fall, to triumph, to recreate life out of life," what custom functions and features might be built into our living machinery?[38] If, in the future, all life is programmable, those with the right knowledge and capability will be in possession of unfathomable power. They could create life, and tweak existing life forms, to do almost anything—be it good or bad.

Which is why a second race—this one involving not just one cell, or one protein, like insulin, but the entire human genome—was an even higher-stakes chase that resulted in an unlikely winner and concerns about who should be allowed write-level permissions to our shared biological code.

| TWO |

# A RACE TO THE STARTING LINE

RESEARCHERS NEEDED A SET OF TOOLS TO ADVANCE THE HYPOTHESIS that if we could decode life, we could reconstruct and repair it—or even redesign it to suit myriad purposes. With the discovery and synthesis of insulin, we built a map, tools, and eventually a computer system and succeeded, but in the process we caused a host of new problems. Making a new discovery turned out to be easier than confronting the political and organizational structures of science itself. And it began with a race with formidable competitors, one side representing the new guard, new science, and private funding, and the other the traditionalists, who favored conservative methods and whose money came from the government.

The contours of that competition started to become apparent when scientists attempted to find an answer to a crucial question that had to be addressed before genes could be sequenced: How closely were genes packed together on a strand of DNA?

In the early 1980s, the US Department of Energy (DOE) and the Office of Science and Technology Policy sponsored a meeting in Utah to discuss genetics and energy. The theme of the meeting descended from a truly awful event and its aftermath. In the years after the United States dropped atomic bombs on Nagasaki and Hiroshima in 1945, the US government conducted an ongoing (and non-voluntary) study on Japanese survivors. Congress had charged the DOE's predecessors, the Atomic Energy Commission and the Energy Research and Development Commission, with studying the effects of radiation.[1] For decades, scientists studied and analyzed the consequences of the chemicals used and the radiation emitted, hoping to understand the genome structure and resulting mutations.

That research was still underway in 1984 when scientists gathered in Utah.[2] Some heavy hitters were in attendance, including biologist David Botstein (MIT), biochemist Ronald Davis (Stanford), and geneticists Mark Skolnick and Ray White (University of Utah).[3] But the conversation took an unexpected turn when geneticist George Church (Harvard) started riffing on the consequences of atomic energy and human evolution. That led him to ideate out loud about the need for a more complete genetic map, which led to a new conversation: theoretically, it was possible to predict the likelihood of two genes being closely linked together based on how often they were separated when DNA split and recombined. This, they reasoned, would allow them to construct genetic linkage maps in humans. Creating a map of the human genome was therefore plausible, even if it wasn't yet technically feasible.

The more Church and the other members of that group thought about it, the more a genome project made sense—but it would require a large-scale effort. Church catalyzed an early effort to explore the idea of the project, which led to a series of other meetings and, eventually, an initiative to sequence the entire human genome.[4] But soon, several federal agencies became involved in a turf war over the proposed scope, funding, and leadership. Some argued that, if mapping the whole human genome (which had never been done before) was to be attempted, surely it needed to be helmed by the National Institutes of Health (NIH)—not the Utah meeting's sponsor, the Department of Energy.[5] Meanwhile, the US National Academy of Sciences had established a special committee to weigh in and advise lawmakers. In 1987, Congress decided that a new organization was needed, one that would be situated in the NIH and called the Human Genome Project (HGP). James Watson, who'd won a Nobel Prize for his work discovering the double-helix structure of DNA and was working at the NIH, appeared before Congress in 1988 to help make the case that breaking into the molecule and decrypting the genome was critically important and must be pursued, even if such a project would require decades of work and billions of dollars.[6]

The NIH and DOE signed a memorandum of understanding to "coordinate research and technical activities related to the human genome," with Watson installed to lead a new Office of Human Genome Research at the NIH and oversee the project.[7] The original plan was to sequence the human genome by 2005—a fifteen-year initiative, with three five-year funding cycles. NIH would get the bulk of the funding, but DOE would play a supporting role.[8]

At that time, a promising young scientist known equally for the speed at which he liked to work and his issues with authority was working at the NIH: John Craig Venter. It would be years before he irked the Joyce estate for publishing a poem inside of a cell without asking for permission first.

Venter grew up in Millbrae, California, a working-class town just west of San Francisco International Airport.[9] Early on, he displayed an unusual appetite for risk. He liked to race airplanes down the runway on his bicycle, and continued to do so even after being told not to—and getting screamed at—by airport guards. His family's modest home sat near railroad tracks, on which he'd sometimes stand while trains sped toward him until, at the very last moment, he'd jump out of the way. In high school, he showed promise in both shop class and biology; a constant tinkerer, he ended up building two speedboats by the time he graduated. He also liked to spend time at the beach, where he surfed whenever the weather cooperated. And often when it didn't.[10]

In 1964, Venter enlisted in the navy to avoid the draft and wound up in San Diego as a medical corpsman—basically a physician's assistant—assigned to the navy hospital. There, he performed spinal taps and liver biopsies in the morning, then headed to the sandy coast of La Jolla to surf in the afternoon. He was eventually dispatched to Vietnam anyway, where he served a bloody tour of duty at the navy hospital in Da Nang during the 1968 Tet Offensive. There were blowups with superior officers throughout. After he came back home, he earned a PhD from the University of California at San Diego, where he studied under the famed biochemist Nathan Kaplan, who had contributed to the Manhattan Project.[11]

Venter began working at NIH in 1984, when most scientists used a laborious process to read entire sequences of each gene they studied. Watching them work, Venter thought back to the time he spent toiling on projects at his workbench, or dealing with the badly wounded patients he treated in Vietnam—both situations where he'd learned to solve problems despite having incomplete information. It made him think that reading gene sequences would be faster if he isolated fragments and then pieced them together, like the way you complete a jigsaw puzzle.

Venter came up with the unorthodox idea to decode bits of genes, rather than entire sequences. He began isolating what are known as expressed sequence tags, or ESTs, which are mRNA strands that have been copied back into DNA using the enzyme reverse transcriptase.[12] These short DNA fragments can provide insights into what genes exist, where they are located in the genome, and whether they are turned on in a particular cell or tissue. Venter used ESTs to identify snippets of previously unknown human genes. If ESTs were jigsaw puzzle pieces, Venter thought he could use custom-built computers to help identify and connect them to see the bigger genetic picture.

This didn't sit well with his fellow researchers, who thought the method was sloppy: skimming, as opposed to the traditional in-depth work they favored. Venter ignored them. By 1991, he'd identified new partial sequences from about 350 human genes, far more than anyone else had; at that point, it represented the fullest knowledge of the human genome.[13] To put that figure into perspective, the human genome contains at least 6.4 billion letters of genetic code, which is roughly the number of letters found in four thousand copies of *Moby Dick*.[14] But 350 was a start, and Venter's new method proved easier, more powerful, and much faster than the traditional way of doing things. Naturally, some scientists seemed threatened. As Venter prepared to submit his research for a peer-reviewed journal, some of his coworkers pleaded with him not to, fearing that their reputations and funding for genome sequencing would be jeopardized. Venter went ahead and submitted his work, sensing that with ultra-powerful computers and sequencers his proposed methodology could scale big and fast—and that a published paper on his techniques would help rally support.[15]

James Watson disapproved of his aggressive young underling.[16] The Human Genome Project was an enormously complex undertaking, a project that he thought would be best handled by many disparate teams. He organized different academic institutions around the country to each sequence DNA, a herculean effort for which he budgeted a staggering $3 billion (roughly $6 billion in today's dollars), which came to him courtesy of several US government agencies and the London-based Wellcome Trust, one of the largest charitable medical foundations in the world.[17,18] Watson and his colleagues drafted an initial five-year plan to set the project's goals. First, it would improve and develop the technology necessary to sequence the human genome by isolating each chromosome and clone fragments to produce clone libraries. These clones would be ordered using both genetic and physical techniques to produce an overlapping set. By the mid-1990s, it would start to sequence these clones and analyze the sequences with computers to identify the genes, and then, eventually, determine which of them were associated with incurable genetic conditions, such as Huntington's disease, fragile X syndrome, and others. Along the way, it would try to develop faster, more automated methods, especially for DNA sequencing.

Watson represented the old guard: the traditionalists who didn't seek out new approaches and were concerned about the speed of Venter's process. But Watson's narrow views weren't just limited to scientific discovery. Before Watson and Francis Crick became famous for discovering DNA's double helix, a brilliant young scientist at King's College London named Rosalind Franklin was using a technique called x-ray crystallography that used x-rays to explore the DNA molecule.[19] She'd been trying to learn how DNA, which was known to have a role in cellular transformation, encoded genetic information. When she beamed x-rays through a crystallized molecule sample, it yielded a characteristic pattern, but of what she wasn't yet sure. A senior academic took her research and showed it to Watson without her knowledge, and the rest of the story you know—he, along with Crick, proposed that a DNA molecule was a double helix, made up of two chains of nucleotides. Not only did Watson refuse to give her credit for the discovery, he later wrote a sexist depiction of her in his book *The Double Helix*. Watson infantilized her, referring to her as "Rosy"—a nickname she never used—and focused solely on her appearance rather than her contributions to science:

> I suspect that in the beginning Maurice hoped that Rosy would calm down. Yet mere inspection suggested that she would not easily bend. By choice she did not emphasize her feminine qualities. Though her features were strong, she was not unattractive and might have been quite stunning had she taken even a mild interest in clothes. This she did not. There was never lipstick to contrast with her straight black hair, while at the age of thirty-one her dresses showed all the imagination of English blue-stocking adolescents. So it was quite easy to imagine her the product of an unsatisfied mother who unduly stressed the desirability of professional careers that could save bright girls from marriages to dull men.[20]

Watson had clear, strong views on women, people of color, and the LGBTQ community, and he certainly didn't think they had any place in the hard sciences or working as researchers. He told a reporter for London's *Sunday Telegraph* in 1997 that if a "gay gene" was ever discovered, women pregnant with fetuses who had it should be granted abortions.[21] In a guest lecture at the University of California at Berkeley, Watson told a group of students that he wouldn't hire a fat person, and he incorrectly told them that there was a genetic link between darker skin and sexual prowess.[22] He said in a 2003 BBC documentary that one useful practical application of genetic research could be to fix what he thought was a scourge of unattractive women: "People say it would be terrible if we made all girls pretty. I think it would be great."[23] In 2007, he told the *London Times* that Africans weren't as smart as Europeans, because of their "breeding": "All [England's] social policies are based on the fact that their intelligence is the same as ours—whereas all the testing says not really."[24] That same year in an *Esquire* interview, he bolstered stereotypes about Jewish people. "Why isn't everyone as intelligent as Ashkenazi Jews?" he asked, intimating that smart, rich people—though perhaps not Jews—should be paid to have more babies.[25] Watson was using even more pointed language by 2019, telling a PBS documentary, "There's a difference on the average between blacks and whites in IQ tests, I would say the difference is genetic."[26]

It isn't hard to understand why Watson felt threatened by Venter, who had long hair for much of his life, was pro-women, and wanted the smartest people around him, whoever they happened to be. All that mattered to Venter was the science.

Venter didn't hide his irritation, either, which caused fissures at the NIH. His process could get the work done faster, he believed, and far more cheaply. Part of why this was happening, he knew, was that the NIH was most comfortable with established structures and approaches. But Venter blamed Watson, who he thought was an inept administrator. The bureaucracy Watson had created, Venter charged, "had become a pointless, annoying and frustrating distraction from science."[27] But Venter also had no patience—or aptitude—for the softer skills of persuasion. Charm and negotiation skills often determine success in large organizations, and his challenging and brusque demeanor won him no friends. In fact, people hated him. He told others, "I was wasting my time, energy and emotion on battling with a group that seemed to have no serious interest in letting an outsider analyze the human genome."[28]

Still, the NIH decided to apply for patents on the gene fragments Venter had identified. This was an important move, because whoever held the patents determined how they could be licensed. Venter wasn't trying to patent the biological material itself—the US Patent and Trademark Office wouldn't have granted such a patent—but rather the code he had sequenced. This move threw Watson into hysterics. He screamed at Bernadine Healy, the director of the NIH, and demanded the institute revoke the funding on the patent effort. (Healy agreed.)[29] But the argument had spilled outside the NIH to Congress. On the Hill, Watson and Venter were summoned to a Senate hearing room in 1991. It was mostly empty. The United States had just started pulling all 540,000 US troops from the Gulf War, and four Los Angeles police officers had been videotaped repeatedly beating Rodney King, so there wasn't a lot of bandwidth for a fight over an obscure topic that few, at the time, could even define.[30,31] A few senators showed up, but none of them evidenced much of an understanding of genomics. They asked rudimentary questions about the project and the patents. At one point, an exasperated Watson, struggling to express his concerns with Venter's work, compared him to a monkey. "It isn't science!" Watson yelled.[32]

In October, the patent office turned down the NIH's application.[33]

Venter, increasingly frustrated with Watson and the NIH, wanted to use some grant money he'd previously received from the institute for DNA sequencing for EST sequencing instead. He asked permission—perhaps he had learned a few things about bureaucracy—but the genome project denied his request and participation. Venter, disgusted, returned the grant money and sent a scathing letter to Watson. Shortly after, he left the NIH, and a venture capitalist named Wallace Steinberg offered to create a company using Venter's EST method. Venter wanted to focus exclusively on basic research and not worry about running a business, so the two reached a compromise. Venter would work with his wife, Claire Fraser, a genome scientist and a specialist in microbial genomics, at The Institute for Genomic Research (TIGR). Steinberg would establish a for-profit company, Human Genome Sciences. The two organizations would work together—Venter's research at TIGR would be used by Human Genome Sciences for commercial development. In the aftermath, Watson was forced to resign as head of the Human Genome Project in 1992, in part because of how he had managed Venter and the patenting debacle.[34] Watson, enraged, receded from public view but continued steering and advising the HGP quietly behind the scenes.

By 1994, the HGP had built enough of the technology and processes to map (but not sequence) the genomes of the fruit fly, yeast, roundworm, and *E. coli*. But it was moving slowly.[35] Meanwhile, Venter and a colleague, Hamilton O. Smith, then at the Johns Hopkins School of Medicine, proposed—you guessed it—speeding things up using another controversial technique: shotgun sequencing. In traditional genome mapping, scientists go through a laborious process of isolating each chromosome, cutting out small pieces of DNA at regular intervals on each one, putting the pieces in order, and then feeding them into sequencing machines that "read" the letters. This process is logical and orderly but slow. It was roughly akin to driving down a long highway in a blizzard while only being able to see a short distance in front of you.[36]

With shotgun sequencing, Smith and Venter would take multiple copies of genomic DNA, shred it, then clone the fragments into bacterial plasmids. Each plasmid would contain a few hundred letters of DNA, which would then be sequenced. Software would read each piece and find overlaps that matched. In this fashion, an entire genome could be assembled. None of the time-consuming ordering of cloned plasmids would be necessary.

It wouldn't be easy: while shotgun sequencing had been used before for smaller projects, it had never been used on something as complex and large as a human genome. Because the DNA had to be shattered at random to create the overlaps, there would be an enormous number of pieces to sequence and reassemble. It would require custom software and computer hardware. But it was an ingenious approach. And it freaked out much of the scientific community.

Smith and Venter applied for a grant from the NIH to shotgun sequence the *Hemophilus influenzae* bacterium, which causes meningitis in children—and then told the NIH they only needed a year to complete the work.[37] This bacterium has 1.8 million letters of code, so this meant identifying and matching roughly 5,000 accurate bits of code every day—including weekends.[38] The NIH review panel gave their proposal a low score and admonished the two on their technique: shotgun sequencing a genome was not only impossible, they said, it was risky. Smith and Venter appealed the decision. But Venter also pushed ahead, assuming that it wasn't worth waiting for the bureaucracy to churn through the appeals process.

A year later, in May 1995, Venter and Smith gave the evening keynote address at the annual meeting of the American Society of Microbiology (ASM) in Washington, DC.[39] Andrew was in the audience, and along with his colleagues Ken Sanderson and Ken Rudd, he had been researching the hybrid (part genetic, part physical) map of the bacterium *Salmonella typhimurium*. Andrew recalls thousands of scientists listening in stunned silence as Venter and Smith announced the complete sequencing of the *H. influenzae* genome, walking them through each step of the process while showing the genome organization in striking detail using computer-generated maps. It was the first time the complete genome of a free-living organism had been sequenced. And then, four years before Steve Jobs began using his famous "one more thing" gimmick, Venter and Smith closed the keynote with the complete genome map of a *second* bacterium, *Mycoplasma genitalium*.

Andrew knew this was a significant announcement. He had assumed that the widely studied and mapped *E. coli* bacterium would be the first to be sequenced. Venter and Smith had come from behind to overtake the entire world of microbiology. Andrew soon decided to leave academia to join the biopharmaceutical company Amgen, which had the technical and financial resources to do Venter-level genome sequencing at scale.

Details of both microbial genomes were published in the prestigious journal *Science* a couple of months later, just as Venter and Smith had planned.[40,41] In a bit of irony that doubtless pleased Venter, the article appeared around the time that he and his colleagues received their final rejection letter from the NIH appeals committee. It informed Smith and Venter that their shotgun method wasn't feasible.

Meanwhile, the distinguished medical geneticist Francis Collins had been installed as the new head of the Human Genome Project after Watson's ouster. But Watson had stuck around—working back channels and frequently sharing his opinions with Collins. TIGR researchers had moved on to other projects, but Venter remained irritated knowing there was another, better way to get the genome completed. He had a point. Internal auditing revealed that, at the rate it was going, only a fraction of the genes would be successfully sequenced by the 2005 deadline. The complex organizational structure that Watson had insisted upon, full of many different groups that had all received grants, weren't all performing. Organizational bloat was killing any momentum, and, perhaps, even the HGP itself.

## THE NEED FOR SPEED

No one would ever mistake Perkin-Elmer Corporation for Exxon or Procter & Gamble. But it was a household name in some very specific households: it owned about 90 percent of the market for the chemicals that scientists needed to sequence DNA. One of its divisions in the 1990s, Applied Biosystems, had been working on a skunkworks project: an automated sequencer called the ABI Prism 3700, which was designed to sequence DNA continuously at high speed by replacing the large flat sequencing gels used at the time with thin, gel-filled capillaries, or tubes.[42]

Andrew had seen a prototype of the device a few months before the ASM meeting when he'd gone to Edmonton, Alberta, to visit Dr. Norm Dovichi, a prominent analytical chemist with expertise in single molecule detection systems, to discuss a collaboration on a new "sequencing by synthesis" concept. Dr. Dovichi had politely listened to Andrew's pitch but said he was too busy. Besides, he was already working on a new sequencer, which he was happy to show Andrew. The prototype had thirty-two capillaries, each equivalent to a "lane" on a slab sequencing gel. (The final version would have ninety-six, the same number of wells on a standard microplate used in robotic lab systems.) Dovichi also shared the performance numbers. Doing some quick calculations, Andrew realized that a single machine would be able to sequence a bacterium in a couple of weeks.[43]

One ABI Prism 3700 wasn't powerful enough to sequence the human genome on its own, but some of the Perkin-Elmer executives believed that several hundred of them working together might be able to decipher human DNA faster than conventional methods. There might be some holes in the resulting code, but it could be thought of as a first pass at the problem; running it through the computers again and again would ensure that all of the missing or garbled bits of code were corrected. That part had the company's finance people concerned, because ultimately, resequencing, as it was known, could take years, not months, and selling the company's chemicals made for much higher profit margins than selling computers.

Perkin-Elmer executives, who knew about Venter and his shotgun technique, thought that pairing their computers with his method could lead to a magnificent breakthrough in genomics. For his part, Venter immediately understood that this approach would speed up the decoding process. So, in 1998, Venter and colleagues walked into the NIH and announced that they were forming a private corporation aimed at sequencing the human genome, one built on Venter's technique and an army of ABI Prism 3700 computers.[44] They made the case for a public-private partnership: with their methodology and computers and the public scientists' more traditional research, they could collaborate and finish decoding the genome before the 2005 deadline, saving a lot of public funding in the process.[45] Venter proposed sharing the data, and once the human genome was published, they'd all share in the glory of accomplishing one of the greatest scientific achievements in human history. No one dared to say it aloud, but if the Nobel Prize was conferred, then they would share in that honor, too.

Collins told Venter he would consider his proposal. But Venter's offer may not have been wholly sincere. He had already contacted the *New York Times* to leak a press release saying that his new company, Celera (Latin for "speed"), would sequence the human genome by 2001, a full four years before the HGP promised it would complete the same task. The release also stated Celera would do this for a fraction of the budget of the public project—under $300 million, less than a tenth of the price tag for the publicly funded, federally led effort. The story the *New York Times* ultimately published intimated that, with Venter's team using a proven method and newfangled supercomputers, the HGP's slower, more traditional method might be made moot.[46] After the article came out, Venter's next stop was a Human Genome Project meeting, where he taunted the attendees, telling them to simply stop working because soon they would be woefully behind.

Venter's trash-talking didn't stop there. After the HGP meeting, there was a press conference to explain what progress had been made. Up at the dais, while sitting next to Collins, Venter told the reporters that the HGP would be better off working toward a more achievable goal: sequencing the genome of a mouse. Perhaps recognizing he'd gone too far, Venter backpedaled, sort of. "The mouse is essential for interpreting the human genome," he offered. After the press conference, Watson—who hadn't been onstage with Collins and Venter—lost it in the lobby, publicly comparing Venter to Hitler.[47] Later, Watson reamed out Collins in front of others, telling him to buck up and be like Churchill, not Chamberlain.[48]

Other HGP scientists weren't just upset about Venter's methods or slashing commentary. They believed that creating a for-profit corporation to conduct such crucial research was simply in bad taste. They also worried that he would not only beat them to the goal, but make it difficult for others to use the research. Technically, Celera could make the genetic code available for anyone to see and use. But without the right computer system, and without deep knowledge of how Venter's method worked, the public database might be wholly incomprehensible. Plus, if anyone wanted real analysis—the important stuff, telling you where things were on the genome—they'd need to pay.

Higher-ups at the Wellcome Trust, the big London-based funder of the HGP, heard all this tumult and understandably grew concerned about a private corporation—especially one run by a renegade from the project it had long backed—suddenly butting in and saying that the public project was wasting a ton of money. Wellcome Trust execs flew to the United States, worried that their considerable charitable donation would be wasted and that the entire project could be in danger. Collins tried to persuade them that the project was fine, saying that Venter was a braggart with a massive ego, and that his approach, using computers and shotgunning, wouldn't work. To top it off, Collins told *USA Today* that Celera would produce the "Cliffs Notes or the Mad Magazine version" of the human genome.[49]

The race was on. Venter said his group would have a working draft of the human genome by 2001 and a completed version by 2003. The HGP had no option. It had to speed things up too. Watson started lobbying Congress for more money for the HGP to purchase its own ABI 3700 machines, each about $300,000. The NIH consolidated the project, concentrating the work into three academic centers: Baylor University, the Massachusetts Institute of Technology (MIT), and Washington University. But doing so meant cutting many other people out. By this point hundreds of scientists had been working for nearly a decade, and their funding was suddenly slashed. It was easy to see why this was happening: all of a sudden, a new, competing effort had reshaped the project.

## CRACKING THE CODE

With a race on, attempts were made to foster cooperation between the international consortium and Celera, but by February 2000, with tensions ratcheting ever higher, talks broke down. Much of that argument was being fought in the press. Venter was incensed that the HGP had leaked a letter it sent to Celera detailing problems with its methods, so he told reporters the HGP had simply resorted to a "low life" response. Meanwhile, a leading member of the HGP said that Celera's plans to sell its proprietary research, along with public genome data, was "a con job."[50]

In March 2000, Venter made a big announcement: Celera, using his techniques and Prism machines, had sequenced the genome of the fruit fly *Drosophila*.[51] That completion validated Venter's assertions all along, as well as Celera's methods. Hundreds of Prisms were now humming along, crunching code, like the enormous present-day server farms you see in movies such as James Bond's *Skyfall*, except that this was all happening years before anyone had invented the phrase "server farm." Venter also said that Celera had started on human DNA and would soon complete a draft of the human genome—roughly 1.2 billion letters of code—along with provisional patents for 6,500 human genes.

Patents anchored a lucrative business model, and whoever owned the patents would reap untold riches. One way to think about what was happening is to envision a game of Monopoly, but one in which moves and privileges were conferred through scientific discovery, not rolls of the dice and random chance. The first team to land on a gene could claim it. Patenting it was like building a house or a hotel: others who either wanted or needed to use that gene had to pay for the privilege (in this case, a licensing fee). Venter's game was underpinned by a simple and distinct strategy: he claimed basically every gene he was able to through provisional patents, intending to later determine which were worth keeping.

In health, medicine, and genomics, patents are important because they give the holder the ability to build lucrative commercial products, such as new drugs for stubborn illnesses. Celera's patents meant that anything at all derived from its genes would belong to the company for seventeen years. But consider what it would mean for different companies to have different rights to the key components of human life—if Celera laid claim to certain genes, while the Human Genome Project did so with other genes, and still other companies or government agencies claimed others. Sequencing the human genome remained crucial, but carving up parts of the map could slow down the process of accruing the necessary information derived from all this work. It would make the collaboration needed to solve knotty health issues with new gene therapies difficult and needlessly expensive. In Monopoly, the red and orange properties are the ones most landed upon, statistically, and therefore the most likely to bring in a lot of rent. If the HGP and Celera were chasing each other around the Monopoly board, each trying to land on as many red and orange properties as they could—that is, sequence and patent the genes that would be the most useful for common gene therapies—that could be a problem.

The scientific establishment agonized over these intellectual property issues. Celera could well patent thousands of human genes, and once it did, who knew what Venter might dream up?

Some of the concerns were existential. If Celera succeeded, it could signal to the outside world that the traditional route of basic research—typically in the hands of big research institutions and government agencies—may not be the best way forward, and that smaller and more agile elite groups could be more effective. Venter's nontraditional methods were a threat.

Celera wasn't just working on experimental new scientific technologies—it was also trying to create a new business model for biotechnology.[52] Venter planned to make the raw data available to the public, but his company would sell software, subscriptions to processed data, and access to its formidable sequencing machine that would make the data usable. At one point, the valuation of Celera was $3.5 billion, which was half a billion dollars more than the budgeted cost of the HGP.

On top of that, the motive to profit from something so universal and innate—our human genome—worried many. Shouldn't the human genome be public domain? Especially since, up until this point, research into it had been funded by taxpayers' money from several countries? Why should just one company reap a financial windfall?

While the science establishment debated, on April 6, 2000, Venter announced that Celera had finished sequencing the necessary DNA and would move on to its "first assembly" of a human genome.[53] It would just be a matter of weeks now before the draft genome was complete—far earlier than anyone had anticipated.[54] This particular genome belonged to one man. (While the man hadn't been identified, some wondered whether it was Venter's own DNA. Venter didn't confirm that it was his—but he didn't deny it, either.) The HGP, on the other hand, had instead used genetic material from several different people.

It was clear now that Celera would beat the public project, but because of all the public funding, all parties agreed that the race to sequence the human genome, and the learnings, discoveries, and new processes that resulted from all that work, would end in a tie: Celera and the HGP would share the credit. But thorny issues remained. Pharmaceutical manufacturers and biotech companies understood that eventually, others—academic researchers, government institutions, startups—would need access to the sequence and to the patents. To create new therapies, they would need the raw genome data and the ability to read it—but they didn't want to have to pay for it. Celera also had two smaller rivals—Human Genome Sciences and Incyte—which had taken massive financial risks in trying to sequence the genome, and they had to earn back what they'd spent.

All these complications reflected the vast implications of this type of research. Whoever controlled access to the genes controlled access to biology's future. In no other area of business would there be a debate over which organization gets to control and profit from how it packaged its research. (The closest analogy, perhaps, is the current debate over access to and control of our individual data, but as of this writing no real action has been taken and none is in sight.) But that underscores why the field of biotechnology—and the world of synthetic biology that was now coming into view—is unlike any other business ever formed by anyone on Earth.

## PEACE TALKS

Concerns over all the issues in play were ultimately taken to the highest levels of government. The resolution came on June 26, 2000, when Venter and Collins, feigning collegiality, joined President Bill Clinton at the White House, with UK prime minister Tony Blair—whose Labour government had been prompted on these matters by Collins and the Wellcome Trust—joining via satellite.[55] Watson was not invited to the dais, although Clinton graciously acknowledged him:

> Nearly two centuries ago, in this room, on this floor, Thomas Jefferson and a trusted aide spread out a magnificent map, a map Jefferson had long prayed he would get to see in his lifetime. The aide was Meriwether Lewis and the map was the product of his courageous expedition across the American frontier, all the way to the Pacific. It was a map that defined the contours and forever expanded the frontiers of our continent and our imagination.

> Today the world is joining us here in the East Room to behold the map of even greater significance. We are here to celebrate the completion of the first survey of the entire human genome. Without a doubt, this is the most important, most wondrous map ever produced by humankind.
>
> It was not even 50 years ago that a young Englishman named Crick and a brash, even younger, American named Watson first discovered the elegant structure of our genetic code. Dr. Watson, the way you announced your discovery in the journal *Nature* was one of the great understatements of all time: "This structure has novel features which are of considerable biological interest."
>
> Thank you, sir.[56]

The president went on to announce that while the race to complete the map had ended, from that point forward public and private research teams would work collaboratively for the benefit of all to complete an error-free final draft of the genome. They would then identify every gene, and ultimately use all of that data to develop new medical treatments.

But the historic announcement ended with a warning. Clinton said that science alone could not be the arbiter of the "ethical, moral and spiritual power" that humanity now possessed. Genetic information should not be used to stigmatize or discriminate against any group. It should never be used to pry open the doors of privacy. Prime Minister Blair went on to emphasize that "we, all of us, share a duty to ensure that the common property of the human genome is used freely for the common good of the whole human race."[57]

Before Clinton congratulated both Venter and Collins and shook their hands, Venter had one last thing to say: "One of the wonderful discoveries that my colleagues and I have made while decoding the DNA of over two dozen species, from viruses to bacteria to plants to insects and now human beings, is that we are all connected to the commonality of the genetic code and evolution. When life is reduced to its very essence, we find that we have many genes in common with every species on earth, and that we are not so different from one another. You may be surprised to learn that your sequences are greater than 90% identical to proteins in other animals."[58]

Venter, who is an atheist, sounded very much like someone who had glimpsed the divine and responded with a newfound humility.[59] He had other reasons to be humble. Despite the magnitude of his genetic sequencing accomplishment, he knew he had only won the race to the starting line. The real game—to design the future of synthetic biology—was only about to begin.

| THREE |

# THE BRICKS OF LIFE

CELLS ARE UNIVERSAL, ELEGANT MACHINES THAT TRANSMIT INFORmation. Although they act like computers—storing, retrieving, and processing data—they look nothing like one. They also act like high-tech, fully automated factories, with departments performing specific tasks to manufacture desired products. These analogies challenge our mental model of how we think about life as a black box: we are aware of (and possibly even control) the inputs, and we can see the outputs. But the inner workings allowing life to be created and its systems manipulated are opaque. If we could manipulate cells—the basic building bricks of life—then we could direct the machines to do our bidding.

If we agree that cells are like wet, biological computers executing commands to produce products and services, then it can be helpful to think of DNA's programming language as digital, except not binary. The computers on your desks and housed inside your smartphones understand 1's and 0's, a two-symbol language (hence, binary) where the symbols represent true (1) or false (0). Those 1's and 0's are strung together, usually in sets of eight, as a byte—a standard unit of digital information. The binary code for the letter A is 01000001. If you wanted to spell out A-M-Y you would need three bytes, with the correct order of 1's and 0's.

The language of DNA uses A-C-T-G, and DNA's version of a byte is a codon, which uses three, rather than eight, positions. For example, ATG codes for the amino acid methionine. When the cell sees that very first ATG, it knows to begin producing protein there. ATG is biology's "hello world."

When the Human Genome Project consortium and Craig Venter's team sequenced the human genome, they identified the roughly twenty thousand genes present in humans. That gave us a better understanding of our source code: the set of instructions detailing the structure, organization, and function of our development and evolution. It was a detailed overview for human cells that would offer insights to identify, treat, and prevent disease. And an opportunity to think much bigger.

Cells contain full copies of the genome, and each cell has the ability to make decisions about its future. A cell can't at the same time be a muscle cell and a skin cell—a choice must be made. Over a cell's lifetime, it has the potential to divide and fork, with each new generation becoming more specialized. But another type of cell—a stem cell—is nonexclusive and universal, and it can divide and replicate again and again, making it an invaluable, renewable resource. Stem cells generate replacement cells for cells damaged during chemotherapy treatments, help the immune system fight blood diseases, and aid in the regeneration of damaged tissue.

In the early part of the twenty-first century, we had a map of the genome—a basic understanding of the location of genes and the distance between them on chromosomes—as well as bold hypotheses about how we might use that knowledge to improve life. What was missing were the tools and the standardized language necessary to program cells. Scientists working at the far edges of this bold new intersection of biology and technology wondered if life wasn't all that mysterious after all, but simply an issue of mechanics—a challenging engineering project waiting to be investigated. But without standardization, a common shared lexicon, and hierarchical systems (parts, devices, and methods), researchers couldn't share their discoveries, disclose patents on newly identified biological configurations, and build on each other's contributions. The basic building blocks of other material realms have been standardized, which is why you can walk into a hardware store and buy a handful of bolts without having to ask whether the thread is the right size. Bolts, screws, nails—and many other things—are all built to particular standards. In engineering, the same is true for metals, polymers, and other materials. In the computer business, hard drives and memory are built with the same basic form factors, so if a drive fails you can order one online, pop open your machine, and replace it. Why wasn't the same true for biological parts?

Think, for a moment, about the magic that this approach would eventually create. In time, there would be dazzling outcomes: a wetware store for standard biological parts and special printers to synthesize molecules. DNA could be reimagined as rewritable data storage, and cells as microscopic production facilities. Among those who understood that this could all happen, there was a new demand for biotechnologists, people who could build a common biological interface. Before we could write new code to control nature—maybe even programs that could direct the future evolution of humanity—someone had to build the hardware store.

As a schoolboy, Marvin Minsky meta-daydreamed: he thought about thinking. He wore round glasses and had a thick, cowlicky head of brown hair, which he tousled while reading the books in his father's library. While other kids in his Bronx, New York, neighborhood were playing stickball, Minsky was devouring the collected works of Freud. He imagined building a replica of the human brain—not an automaton, but a true cognitive machine. One that matched our raw computational power and also had our ability to create, imagine, and feel. He got to explore this idea as a likable, affable student at the prestigious Bronx High School of Science, whose alumni included a number of Nobel Prize winners, and later at the Phillips Academy in Andover, Massachusetts, where famed geneticist George Church would eventually study.

Minsky enrolled at Harvard intending to study math in 1946, but soon his academic path became unclear. He studied math and physics—and psychology, language, even music composition with Irving Fine, whose acolytes had included Aaron Copeland and Leonard Bernstein. While an undergraduate, Minsky also ran his own laboratories. This was a rare feat for an underclassman, and, rarer still, the work they did cut across disciplines. One was in biology, the other psychology. Yet he spent most of his research time attempting to work out the human mind: why it worked, where thoughts originated, how it commanded other functions in the body, its interaction with organs and cells, and whether or not we really have free will. Increasingly, his insatiable curiosity homed in on what he thought were the three most interesting problems in the world: genetics, physics, and human intelligence.[1,2,3]

By the mid-1950s, Minsky had completed a doctorate degree in mathematics, but he couldn't shake his fixation on figuring out, on an elemental level, how the brain worked. In 1956, he and a few friends, John McCarthy (a mathematician), Claude Shannon (a mathematician and a cryptographer at Bell Labs), and Nathaniel Rochester (a computer scientist at IBM), proposed a two-month workshop to explore the human mind and the issue of whether machines might someday think, just like people do. Together with an interdisciplinary group of independent researchers spanning many fields—computer science, psychology, mathematics, neuroscience, and physics—they spent two summer months at Dartmouth exploring the connection between mind and machine. By the end, they had proposed a new field of study, which they called "artificial intelligence," the same AI we know today.[4]

Minsky and his fellow researchers that summer were not the first to investigate human thought and why our cells function seemingly autonomously. Plato and Socrates wondered what it could mean to "know thyself" in ancient Greece, as they, too, attempted to reverse-engineer thought and identity. Aristotle developed syllogistic logic, as well as our first formal system of deductive reasoning, which led Euclid to create the first mathematical algorithm when he figured out how to find the greatest common divisor of two numbers. While seemingly unrelated to cells and the human genome, this work laid the foundation for critical ideas in synthetic biology: that certain physical systems can operate as a set of logical rules, and that human thinking itself might be a symbolic system, a set of codes and rules.

These early ideas in philosophy and mathematics resulted in hundreds of years of inquiry for scientists seeking to discover how our minds are connected to our bodies—the container for the millions of cells, each operating individually and as complex systems, that are making decisions and keeping us alive. Why did our biology work like the exquisite grandfather clock to which it was so often compared? The French mathematician and philosopher René Descartes questioned consciousness and how we could even go about verifying our thoughts as real. His *Meditations on First Philosophy* offered a thought experiment, asking readers to imagine a demon purposely creating an illusion of their world. If your physical, sensory experience of swimming in a lake was nothing more than the demon's construct, then you couldn't really *know* that you were swimming. But in Descartes's view, if you had self-awareness of your own existence, then you had met the criteria for knowledge. "I am, I exist, whenever it is uttered from me, or conceived by the mind, necessarily is true," he wrote. In other words, the fact of our existence is beyond doubt, even if there is a deceptive demon in our midst. Or, in the more famous quote from Descartes, "I think, therefore I am." Later, in his *Treatise of Man*, Descartes argued that humans could probably make an automaton—in this case, a small animal—that would be indistinguishable from the real thing. But even if we someday created a mechanized human, it would never pass as real, Descartes argued, because it would lack a mind and therefore a soul. Unlike humans, a machine could never meet the criteria for knowledge: it could never have self-awareness as we do. For Descartes, consciousness occurred internally. The soul was the ghost in our bodies' machines.

Charles Darwin returned from his voyage around the world on the HMS *Beagle* in 1836. Along the way, among many other things, he'd discovered massive skull bones, the remains of ancient ground sloths and other fossils from Earth's primeval environment, and on the Galapagos Islands he saw different species of finches and giant tortoises and marveled at how the creatures varied slightly from island to island. In the years after his return, Darwin became fixated on the cycle of life—birth and extinction—wondering about hierarchies and heritability, and he developed a theory that all species survived through a "natural selection" process. Living things that successfully adapted or evolved to meet the challenges of their environments went on to reproduce and thrive; those that couldn't adapt or evolve didn't, and eventually died off. This was true of birds and tortoises, of the fossilized animals he'd encountered, of ferns and trees, and of human beings. All life, he posited, was born of common ancestors and evolved over very long periods of time. There was no divine intervention; no day when God created all the animals that inhabit the Earth, and then the first man and woman, as the Victorians believed. God was not a creator, but a strong, ingrained tribal survival strategy we developed as part of our own natural selection process.[5]

While Darwin was finding a common biological language linking all living things together, English mathematician Ada Lovelace and scientist Charles Babbage were trying to mimic part of our biology—cognitive function—through engineering. They created a machine called the "Difference Engine," which tabulated numbers, in the 1820s, and then went on to postulate a more advanced "Analytical Engine," which used a series of predetermined steps to solve mathematical problems. In the footnotes of a scientific paper that Lovelace translated in 1842, she conceived of an even more complex system that could follow instructions and produce music and art, effectively writing the first, if conceptual, computer program. The leap from theoretical thinking machines to computers that began to mimic human thought happened in the 1930s with the publication of two seminal papers: mathematician Alan Turing's "On Computable Numbers, with an Application to the *Entscheidungsproblem*," and Claude Shannon's "A Symbolic Analysis of Switching and Relay Circuits." (Shannon went on to push the boundaries of math, engineering, and human thought with Minsky during that magical summer at Dartmouth.)[6]

All of this background is important, because it shows just how long we've been thinking about what life is—from our biological machinery to our minds, and how the two work together.

By the mid-1960s, Minsky had published seminal papers on AI that addressed the future, looming challenges of self-aware machines. He founded MIT's AI Laboratory and began researching multifarious human-machine problems, such as how to teach computers to reproduce and understand language. One of his most promising students—Tom Knight—wasn't technically a undergraduate. Knight was a high schooler from Wakefield, Massachusetts, a quiet old town with enormous forests and lakes that was a twenty-minute drive north from the MIT campus. Knight spent the summers of his junior and senior years working for Minsky while taking MIT courses in computer programming and organic chemistry. He matriculated to MIT after he graduated high school, but, like Minsky, he had a difficult time finding the right academic field. There was no computer science major—at that point, the field was still in the process of being formed—and back then universities still discouraged hybrid courses of study.[7]

So instead, Knight devoted his energy, as Minsky did, to enabling machines to think. Knight, whose full beard, clean-shaven upper lip, thick swoosh of dark hair, and glasses made him look like a nerdy Mennonite, quickly became a cult figure on campus. In 1967, he wrote the original kernel for an operating system that tracked the time users spent on computers—a crucial task, because, back then, computers were only found at universities and in government labs, and only one person could use them at a time. He was on the team that helped build ARPANET, which later became the NSF Net, and eventually the internet that we use today. In the 1970s, he designed one of the first semiconductor memory-based bitmap displays and a bitmap-oriented printer. In 1978, he and an MIT colleague, Richard Greenblatt, tried to build a simpler computer, one that could eventually be used by anyone, not just trained programmers. They succeeded in building that machine, but they failed at building the business for it. But both went on to start other companies that manufactured computers: Greenblatt formed Lisp Machines, and Knight started Symbolics, which in 1985 became the first dot-com domain ever registered.[8]

Along the way, Knight kept racking up dozens of patents for his contributions in computer science and electrical engineering, and, as part of his doctoral thesis, which he completed in 1983, he figured out how to engineer integrated circuits. Moore's Law, the theory that the number of possible transistors that could be placed on an integrated circuit board for the same price would double every eighteen to twenty-four months, was still holding, and Knight predicted that in the near future, fitting more transistors on an integrated circuit would become physically impossible using traditional engineering methods. Knight figured that at some point, the doubling would result in nanometers. Ten nanometers was just sixty atoms across. It would then become statistically improbable (but not technically impossible) that systems that worked properly could be engineered.

Knight, recalling the organic chemistry courses he took as a teen at MIT, wondered if molecules could be coaxed into self-assembly for the purpose of building a better computer chip. He immersed himself in his old biology books as a refresher, and then found more modern books about simple organisms. He read the works of the biophysicist Harold Morowitz, who argued that "all of biological process begins with the capture of solar photons and terminates with the flow of heat to the environment," and who also explained the thermodynamics required to make a perfect pizza.[9] In fact, a proven technology was already widely used, and it was capable of moving atoms around to precise specifications: chemistry.[10]

While remaining an active professor at MIT, Knight re-enrolled there as a student in 1995 and took the graduate core courses in biology.[11] He wanted to explore the flow of biological information and the physical architecture needed to enable it. The fundamental unit of heredity information is a gene; DNA is both the code and the storage; and a gene's expression permits or halts protein production. Proteins are the molecules that execute commands within cells; proteins called enzymes catalyze chemical reactions. This linked series of reactions form a metabolic pathway, where the product of one reaction becomes the input for the next. Which genes are expressed and when—on or off—represents the flow of biological information. To biologists, this is what's known as the central dogma. To Knight, it represented untapped potential.

In your high school biology class, your teacher might have asked you to dissect a frog and its organs, note what you saw, and then put the creature back together. (At least, that was Amy's experience.) Similarly, in the mid-1990s, the nascent field of biotechnology was still focusing on deconstruction and observation. Knight didn't want to work on the molecular analog to frog cadavers. Nor did he simply want to develop clones of genes, cells, and tissues, which was a core activity in biotech at that time. It wouldn't get him closer to answering the questions he was exploring: Are humans just squishy machines? Could cells be programmed like computers? Could computers be redesigned from biological parts?

Answers would require deep exploration, and it would mean confronting some deeply held beliefs in academia about keeping research siloed. Computer science, robotics, and artificial intelligence alone could not lead to answers. Nor would traditional approaches to biology or chemistry. The engineering approach used to understand computing—around machines, information transfer, connectivity, networking, and autonomous decisions—could be applied to the machinery of cells. But researchers would then need to embrace biology's complexity and variability.

A new toolset and an attempt at standardization could unlock different approaches. A series of biological parts (bioparts) could encode biological functions; devices made from bioparts could encode defined functions; and systems would perform tasks. The far-reaching influence of Minsky's interdisciplinary group at Dartmouth—which had led to the creation of artificial intelligence as a field of study, the next generation of computing, and breakthroughs in myriad areas of research—was in the back of Knight's mind as he convened his own summer workshop in 1995 on Cape Cod. There, he gathered a collective of scientists to focus on merging engineering, computer science, biology, and chemistry in an unprecedented attempt to define biology as a technology platform, the genome as code, and living matter as programmable objects. He called it cellular computing.[12]

The following academic year, he convinced MIT to fund a molecular biology lab within the university's famous computer science lab. But he quickly found another hurdle standing in the way of computable cells. Every experiment the lab attempted required building the piece of DNA the team needed, often from scratch. That required its own sets of experiments, and wasted a lot of time.

## THE SYNTHETIC BIOLOGY ERA

At the time Tom Knight was forging a new field of science, a new strain of drug-resistant malaria was ravaging Africa. Just one bite from an infected mosquito could lead to serious illness or death; indeed, 200 million people bitten by mosquitoes were becoming infected each year, and nearly 2 million of these people died.[13] The medication most commonly used, chloroquine, was quickly becoming useless, because a new strain had developed a resistance to the drug. This new strain was taking over, even in places where chloroquine had never been used. There were other forms of the disease being spread by the large, quickly reproducing mosquito population as well. What further complicated matters was malaria itself. It's very stealthy, often seeming like other diseases. Its symptoms range from chills and sweating, to diarrhea and headaches, to feeling just a bit out of sorts. Sometimes it sits dormant for a year and then suddenly appears, seemingly out of nowhere.[14]

Like many other parts of the world, China had been dealing with malaria for millennia. Ancient herbalists relied on something called *qing-hao*—sweet wormwood—an aromatic herb with a relatively short growing season. Its leaves look like dense green carrot tops. A Chinese physician, Ge Hang, in his *Handbook of Prescriptions for Emergency Treatments*, recommended treating the symptoms of malaria with sweet wormwood in 340 CE*—even before the illness was named or its transmission understood. During the Vietnam War, mosquitoes were an unexpected complication for all sides. Long before chloroquine became effectively useless in Africa, drug-resistant strains of malaria were spreading in Southeast Asia, and the disease overwhelmed troops. So China launched a secret government project aimed at finding an effective malaria treatment for China's North Vietnamese allies. A Chinese phytochemist—a scientist specializing in the chemistry of plants—named Tu Youyou became involved. She and her team of researchers pored over ancient medical texts, including Ge Hang's handbook, and used their knowledge of traditional Chinese medicine to identify 640 plants and 2,000 potential remedies for malaria. Eventually, they found a good candidate in a very old herb: Ge Hang's sweet wormwood. By 1972, the team had isolated an active compound found in a nontoxic extract, which they named *quighaosu*, or artemisinin. China's scientists were prohibited at the time from sharing their discoveries with the outside world, so she couldn't publish her findings. When her discovery was finally made public in the early 1980s, it became the basis for new and better treatments. Tu received a Nobel Prize belatedly recognizing her discoveries in 2015.[15]

The problem with sweet wormwood is that the growing conditions have to be just right. It needs full sun and well-drained soil, and it won't thrive in wet conditions. To meet the new demand for the herb, commercial production of it was ramped up in China, Southeast Asia, and

* CE stands for "Common Era" and is an alternative to AD, or Anno Domini, which in Latin means "in the year of the Lord."

Africa. But it was such a finicky plant that the quality, supply, and cost were inconsistent. Quantities were difficult to predict in advance, which made the global supply chain for artemisinin unstable. By the early 1990s, artemisinin, taken in combination with other drugs, was the only consistently effective treatment for malaria, but the demand for the drug was far outpacing farmers' ability to grow and harvest the herb.

While Knight was contemplating programmable cells in Cambridge, a few thousand miles away, at the University of California in Berkeley, a biochemical engineer and assistant professor named Jay Keasling was also mulling the intersection of engineering, computer science, chemistry, and molecular biology. His senior colleagues advised him to stick with the established research underway, such as predicting the outcomes of inserting new genes into cells. But Keasling was fascinated instead with the prospect of building a new set of tools. He was especially interested in metabolic pathways, and in new ways to rewire cells so that future organisms could do something beyond what they'd evolved to do. He'd already developed a sort of biological rheostat—a switch that could control the current of genes, not unlike the dimmer switch you might have in your bedroom to gradually turn up or down your lights. He applied this approach to various metabolic pathways from different organisms, trying to design new biological circuits in an attempt to modulate outcomes—or even invent new ones.

Keasling and his team started looking at terpenoids, byproducts created by the metabolic pathways in plants.[16] A peony's distinctive, sweet smell; the bright yellow of turmeric and mustard seed; the sticky residue known as pitch (used to waterproof boats); the beneficial compounds in cannabis—these are all examples of terpenoids. And they weren't easy to re-create in 1995. Some scientists had attempted to graft different plant genes into microbes to produce terpenoids from scratch, but the process was costly and didn't yield much. So Keasling and his team thought they'd explore another metabolic pathway.[17] Rather than working on closely related plants, they tried something quite different: yeast.

If you've ever made pizza dough or baked a loaf of bread, you'll undoubtedly remember that magical step of the process when you add granules of brown yeast to warm water and sugar, agitate it just a bit, and then watch it form a foam-topped slurry. Yeast is a single-celled microorganism that feeds on sugar (and expels carbon dioxide in the process, which is what makes dough rise). Keasling and his team inserted the metabolic pathway into colonies, which created an abundant reaction. Then, they added another pathway from a plant, which they hoped would result in a by-product close to a terpenoid. But which one to reproduce? A carotenoid, which makes daffodils bright yellow and tomatoes red? Or a menthol or camphor?

One of the team members was familiar with Tu's research on sweet wormwood and artemisinin. While it seemed promising, at that time it hadn't yet been picked up by major pharmaceutical companies, which preferred to continue marketing their existing malaria treatments. Chloroquine was cheap and highly profitable: it was offered at roughly 10 cents per treatment. Adding artemisinin could drive the price up to $2.40 legally, and as high as $27 on the black market.[18] But Keasling's team realized that none of the issues with naturally harvested artemisinin would matter if it could be manufactured synthetically. They set the rheostat on high to brew a precursor molecule called farnesyl pyrophosphate (FPP), and turned down the genes that convert FPP into a material that forms the scaffolding of yeast cell walls. Then, they used a sweet wormwood gene that converts FPP into artemisinic acid, inserted it into the yeast genome—and the microbes reproduced. It wasn't artemisinin itself, but it was a good start.

Keasling and Knight, on opposite coasts, were coming to similar conclusions. If they reimagined cells as programmable computers or factories, and if they had the tools to direct the flow of information, humankind would no longer need to simply accept the inevitability of natural selection. The modern synthetic biology era had begun.

BOTH KEASLING AND Knight knew that for synthetic biology to move from their somewhat fringy labs to mainstream academia, and ultimately to the outside world, sturdier foundations would be required for their new hybrid discipline. There needed to be more students in the pipeline interested in the intersection of engineering, computer science, and biology. Some way to standardize key components, too. Continuously building pieces of DNA from scratch, and then discovering various metabolic pathways, was a tedious, time-consuming process, and that prevented teams from doing the more interesting—and more societally beneficial—work of fabricating new organisms. It would be much better to have that biological hardware store, and its readily available repository of bioparts, devices, and systems, in operation.

Both Keasling's and Knight's teams were embarking on bold new initiatives by 2002. Drew Endy, a biotechnologist and biochemical engineer on Knight's team, was working on standardizing DNA assembly, so that parts could be combined into working devices, metabolic pathways, and systems. Knight was a lifelong Lego fan, and he drew inspiration from those iconic bricks and parts. He envisioned a toolset of sorts called BioBricks, a set of biological parts with standardized sequences that could be assembled and reassembled as needed. But to make it work, he and his team needed to learn what others might do with such a system, and how much interest there might be on the MIT campus, where the project was based and initially intended to be used.[19] They considered teaching a synthetic biology course, but the ideas were still so new that no pedagogical ecosystem for it existed: no textbooks, no standard curriculum, and very few real-world experiments to use as case studies.

They set up a course anyway, and modeled it on a famous class taught at MIT during the summer of 1978 by Lynn Conway, a computer scientist and electrical engineer. Along with one of her colleagues, Caltech professor Carver Mead, Conway had created a course on a new kind of microchip design while the two were still in the process of inventing it. Conway and Mead, along with their students, rapidly prototyped circuits and transmitted their designs over ARPANET to a chip foundry in California. The project, partially funded with a grant from the Defense Advanced Research Projects Agency (DARPA), resulted in copies of working chips a month later. That class had revolutionized how chips were designed and used, proving that an alternative infrastructure for chip design and manufacturing could work at scale; by doing so, it set the evolutionary path of how chips were designed, the machines that used them, and the business ecosystems that required the machines.[20]

MIT has a winter study offering called the Independent Activities Period (IAP) when students and faculty can take short courses on a wide variety of unusual, experimental topics. Endy and Knight, recalling the success of Conway's class and the revolution it sparked, decided to do something similar by offering an IAP course on BioBricks and synthetic biology. Their plan was to have students design and build a DNA circuit, and then to have it sent over the modern commercial internet to a Seattle-based foundry for printing.[21] One of the first projects from the class, which became known as a repressilator, was designed by students Mike Elowitz and Stanislas Leibler in 2000.[22] The repressilator was to be a small circuit engineered into *E. coli* that had three repressor genes. For convenience, let's call them A, B, and C. These repressor genes were to be connected together in a feedback loop. The protein that gene A made would repress the production of protein from gene B. The protein that gene B made would repress the production of protein from gene C. To close the loop, the protein from gene C would repress the production of protein from gene A. Protein C would also repress another gene, one that encoded for green fluorescent protein (GFP). This produces a visual signal of what's happening with the system. When the system started, each protein would be produced, including GFP, so the cell would be green. As the various repressors started to work on each gene, it would create a cycle. C would repress A (and GFP, so the cell would dim), which would increase B, which would repress C (so the cell would brighten). As the system dynamics stabilized, the amount of GFP produced would oscillate between low and high. Basically, the cells would slowly blink on and off.

Or, at least, that was what should have happened. When Endy received the sequences back from the Seattle foundry, they didn't exactly work as intended. The first generation of BioBricks fit together as intended, but they didn't produce the desired results. The issue was that they simply weren't working with enough genetic material. They were trying to build a Master-level Lego project with a fraction of the pieces required.

Luckily, Endy and Knight's 2003 IAP course was popular, so they divided the class into teams, and each team worked on a different project.[23] They provided each team with some standard preexisting parts, along with funds to synthesize about five thousand base pairs of new DNA, but the designs the teams came up with were all too long. They needed a bigger library of standardized parts and a registry to track their data.

In the summer of 2004, Endy and Knight organized the first International Meeting on Synthetic Biology, SB1.0, at MIT.[24] The three-day conference brought together researchers interested in how to design and build standardized biological systems—and what that could mean for society. It ultimately led to the founding of the International Genetically Engineered Machine, or iGEM, competition, as well as the first Registry of Standard Biological Parts, which was housed at MIT. Andrew was closely following the developments of this group and wanted to get involved. He helped organize a smaller meeting with Endy and others at a new genome center in Oklahoma, worked with the Toronto iGEM team in 2005, and became an iGEM ambassador in 2006, helping grow the program from thirteen to thirty-nine teams.

In addition to parts, a standard method of measurement would be required—BioBricks needed to be described using data, which would allow researchers to build a wider knowledge base for others to use. Soon, the community of researchers began building systems to catalog descriptions of the bioparts, devices, and systems needed, codified for a peer-reviewed database. A wider community would begin to standardize a special computer language, which was called SBOL, for Synthetic Biology Open Language. That would make the data machine readable and easily integrated into different software tools.

One of Endy's friends, physicist Rob Carlson, had been tracking the rates at which different biotechnologies were improving. He was especially interested in DNA synthesis. He calculated that, by 2010, a lab worker would be able to synthesize several human genomes, from scratch, every single day at a cost of just 10 to 12 cents per base pair. Given how long it had taken the Human Genome Project and Craig Venter's teams to do their work—to say nothing of the ten-figure sums it had cost—this was an audacious statement. But Carlson backed it up with data and what appeared to be reasonable models. Endy, Knight, Keasling, and others saw a future in which engineering biological systems was possible at scale. A new revolution was set to begin.

## COMPANY NUMBER ONE

Meanwhile, in 2002 Keasling's research team managed to produce a few micrograms of artemisinin using its metabolic pathway technology. They published the results in the prestigious journal *Nature Biotechnology*, which further detailed this new process for inserting genes into *E. coli*.[25] But Keasling knew that success in the lab wouldn't yet save millions of people from malaria. He secured a $42.6 million grant from the Bill and Melinda Gates Foundation to explore how to ramp up production, and in 2003 formed a company called Amyris Biotechnologies—the first synthetic biology company intended to produce products—with the express aim of creating an artemisinin therapy available to all.[26] Keasling, along with Endy, also launched a counterpart to the iGEMs called BIOFAB, with the goal of professionally developing and cataloging new and existing biological parts.[27]

In 2008, Amyris, which had developed artemisinin in a lab but hadn't built the facilities to make it commercially viable at scale, agreed to give a royalty-free license to French pharmaceutical giant Sanofi-Aventis, which promised to manufacture and distribute artemisinin using Keasling's synthetic biology technique for what it called a "no profit, no loss" margin of $350–$400 per kilogram. The drug was expected to hit shelves before 2012. But this development, and its potential to save millions from malaria, made global news, and once farmers got wind of how valuable sweet wormwood was, they started planting tens of thousands of hectares of it. The market in Asia quickly became oversaturated. There was so much sweet wormwood that the price collapsed, going from $1,100 per kilogram to less than $200.[28] Making things worse, Sanofi sold other products to markets where sweet wormwood was grown naturally, including China. Local pharmaceutical companies were reluctant to do business with Sanofi because they now saw it as a competitor.

Still, consider what Amyris had accomplished. Monsanto, the agricultural mega-corporation, has a global customer base and at the time had a billion-dollar R&D budget. By that point, it had only stacked the genetic code of a corn strain with eight new genes. Amyris, working out of a comparatively small lab near Berkeley, in Emeryville, California, and with far less budget, had managed to engineer thirteen new genes into yeast.

Legendary Silicon Valley venture capitalists came calling. John Doerr of Kleiner Perkins Caufield & Byers wanted to fund synthetic biology applications at Amyris. So did Vinod Khosla, the cofounder of Sun Microsystems, and Geoff Duyk, of TPG Biotech. They weren't so excited about malaria, and they weren't interested in open source databases or free licenses. Instead, they saw a path to disrupt the petroleum industry with these new innovations. They wanted the company to focus on using synthetic biology to create biofuels. When he founded Amyris, Keasling invited four postdoctoral students to see what they could do with artemisinin. But none of them had the background or experience to lead the kind of company dreamed up by its investors, who wanted a battle-tested corporate executive in the role of CEO.[29]

As the story often goes, the investors got their way. Amyris became a biotech company that focused on fuels and chemicals, which could all be modified using Keasling's yeast technology, and it hired John Melo, who had been president of the United States' fuel operations at British Petroleum, as CEO. Almost immediately there were problems. Melo and his newly installed senior managers wanted to track productivity metrics on a team of seasoned researchers, who weren't accustomed to their R&D output being scheduled, and who knew that breakthroughs happened when they happened. He also created a subsidiary in Brazil to produce cheap, plentiful sugar—which the yeast needed to manufacture their biofuel products.

Melo and Amyris's board decided the company should go public and raise hundreds of millions of dollars. They launched an IPO road show in 2010, describing a future demand for fuel and the prospect of cheaply synthesizing it out of yeast. Enthusiasm—and estimates of how much fuel could be produced using synthetic biology—kept growing. The company debuted on the Nasdaq in September of that same year for $16 a share, raising an initial $85 million at a valuation of $680 million. During an interview at Nasdaq, Melo beamed, saying, "Brazil is like the Saudi Arabia of biomass."[30]

Melo promised that by 2012, Amyris could produce 50 million liters of farnesene, a chemical compound that could be converted into a replacement for diesel fuel, which could be used in cars and even jets without harming the environment. It would be the end of natural fuel and the beginning of the biofuel era, they promised. It was yet another familiar, self-imposed deadline: Sanofi said it would start making synthetic artemisinin at scale for the masses that same year. But 2012 came and went. At the end of the year, there was no new drug available, and no cheap alternative to diesel fuel on tap, either.[31]

## SCIENCE VS. BUSINESS

The problem with any novel, complex technology is that it simultaneously builds expectations and raises unrealistic timelines for new products and services. Sequencing the human genome was a project with a clear start and finish. Unlocking synthetic biology revealed thrilling possibilities about how life might be directed and evolve, but it was still basic research, only just emerging as a new scientific discipline.

The promise of Amyris was its novel approach to synthesizing biological materials, but the company, and the science, needed time to mature. Keasling's original concept was that a biologist would someday design virtual genetic code on a computer, test it using algorithmic models, and eventually print the winning combination—whereby robots and machines would then automatically produce the organisms needed. Could that someday be farnesene? Sure, some day. But the mere promise of turning sugar into fuel was itself a sort of chemical reagent, causing a chain of reactions among investors: as their enthusiasm ratcheted up, their patience tanked, and soon they were jacked up on the biofuel economy like teenagers on Red Bull and Skittles.

But the bigger point is this: viewing this new era of science through the lens of tech startups served only to raise and then crush expectations. Because, while synthetic biology is derived from engineering, computer science, and AI, it is something fundamentally different—it's a technology built out of biology. It is a fast-growing field but hardly mature. Investors failed to calibrate the speed of synthetic biology correctly, so they failed to grasp that the field is still in its picks-and-shovels phase. Yes, there will be incredible products that will change our existence on this planet—and perhaps off-planet living, too—but investment was better suited to building and improving the hardware store ecosystem: the materials and the inventory, the supply chain and all the pieces of the value chain.

Artemisinin is hailed as synthetic biology's first successful product, even though the business created to manufacture and sell it was a failure. The fall of Amyris is a cautionary tale. Keasling once estimated that it took about 150 person-years of work to get to the metabolic pathway discoveries his team made.[32] The next steps should have been seeding all the foundational businesses. Instead, investors, journalists, and frankly, a lot of scientists, chided synthetic biology and questioned whether all the hype was commensurate with the value it brought to academic research and to the marketplace.

Compared to scientific discoveries in other fields, such as physics, aerospace, and chemistry, what the foreparents of synthetic biology—Keasling, Knight, Endy, Venter, Church, Collins, and yes, even Watson—had managed to accomplish in less than two decades was truly remarkable. Astonishing. Breathtaking. That government agencies and philanthropic institutions weren't lined up asking to fund the next iteration of their work tells us something about the challenges of establishing new fields of science. Artificial intelligence, which has a strong footing today, with an enormous, global ecosystem, endured what's widely known as the AI Winter in the 1980s, when investments and bold promises from the 1960s and 1970s failed to produce speculative commercial products—such as computers that could automatically translate languages in real time—or government intelligence tools, as was the case.[33]

This new era of science is everything traditionalists tend to abhor: speed, novel techniques, and a blending of disparate disciplines. It flips the script on our cherished beliefs about life and its origins. The very existence of synthetic biology challenges the status quo, and that makes a lot of people uncomfortable. Many scientists criticize BioBricks, arguing that the methodology of its assembly is too slow, or too simplistic, or too . . . *Legos.* By then, Knight was using his Lego analogy regularly to describe the concept of biological parts and the hardware store to non-biologists because it was a relatable analogy. But one group took the explanation literally, arguing that Legos might be fun for children to play with, but no one would actually live in a Lego home.

If there is a lesson to be learned, it is that science demands an open mind and a lot of patience. We may not produce biofuels at a massive scale yet, but we do have a tremendous foundation for our future. In the coming chapters we will detail all the ways—at times both piecemeal and profound—in which our lives will begin to change in the next decades because of synthetic biology. We already have brand-new bodies of research and scholarship upon which others can build. A new lexicon to describe biological tools and processes. Bioparts, devices and systems, and blueprints for new ways to store and access biological data. A new coding language to make DNA machine readable. The beginnings of a hardware store, with everything we need to design the future of life.

| FOUR |

# GOD, A CHURCH, AND A (MOSTLY) WOOLLY MAMMOTH

A MAN, HIS TEENAGE MISTRESS, AND HER STEPSISTER CHECKED INTO the Hôtel d'Angleterre, a modest house near the shore of Lake Geneva in Switzerland, in May 1816. The man, a poet, had gotten his younger lover—Mary—pregnant two years earlier. But a few days after she welcomed that baby, before she'd even settled on a name, Mary found him dead. Naive, and terrified of what might have caused the baby's death, she worried that the milk from her engorged breasts might have been poisoned, and that it might also poison her. She endured dreams that gave her glimpses of an alternate, happier reality. The nightmares began again the moment her eyes opened. In her diary, she wrote, "Dream that my little baby came to life again; that it had only been cold, and that we rubbed it before the fire, and it lived."[1]

By the time Mary arrived at Lake Geneva, she had been pregnant again, and had a healthy infant son. But she was still devastated by her loss. One night, another poet among her circle of friends who had come to stay at the lake suggested they all write ghost stories. A prolific writer, she began channeling her pain into a new work of fiction about a lifeless body that had been pieced together from cadavers by a Swiss scientist named Victor, who then brought it to life. She didn't give the monster himself a name. She published the book anonymously, but today, we know that Mary Shelley was the author of *Frankenstein; or, the Modern Prometheus.*

In one passage of the book, Victor exclaims, "So much has been done... more, far more, will I achieve; treading in the steps already marked, I will pioneer a new way, explore unknown powers, and unfold to the world the deepest mysteries of creation. . . . What had been the study and desire of the wisest men since the creation of the world was now within my grasp." *Frankenstein* endures because it forces us to consider our own origins, as well as our long-standing struggles to understand creation and control. Such questions have fascinated us forever: What is this thing, life? How did it begin? Does it really end? Can we command it to our liking?

Nearly every culture answers questions about life's origins with a set of characters and a story. In Greek mythology, there was Chaos—a nothingness. From that void emerged Gaia, the earth, who gave birth to Uranus, the sky. From them came the Titans, the one-eyed Cyclopes, hundred-handed creatures, the gods (Hestia, Demeter, Zeus), and, eventually, humans. Ancient Sumerians believed in a goddess called Nammu, a mother figure who gave birth to the heavens and to the Earth, and then the plants and animals, and eventually humans. In the tradition of the Lakota, there was a world before this world. The humans there did not behave themselves, so the Great Spirit flooded the Earth, and only Kangi, a crow, survived. Three more animals were sent to retrieve mud, and the Great Spirit shaped it into land, distributed animals all around the Earth, and eventually shaped men and women out of red, white, black, and yellow mud. In Christian stories, God created a desolate, formless space, then light, the sky, land, animals, and finally, Adam and Eve, who were granted dominion over all living things and became parents of the human race.

All these stories were written long before we understood much about biology, natural selection, or how life evolves. The Book of Genesis encompasses dramatic tales of a world in peril, a family that struggles to have children, and a quest to build a land for the future. Its authors were innocent of Darwin's observations about natural selection, or the fundamental laws of genetics that Gregor Mendel outlined centuries later. (The genetic sequences of some of the Bible's famous families—those of Sarah,

Rebecca, and Rachel—might have explained why some of them struggled to get and stay pregnant.)

The Scottish philosopher David Hume observed that our universal creation myths exist because we need stories of cause and effect to make sense of the world around us, and because society functions better when rules have context.[2] What happens now, as synthetic biology breaks our paradigms, forces us to reconsider the rules, and challenges our origin stories? The future of life is being imagined, designed, and manufactured by scientists today in hundreds of laboratories—including one where a beloved researcher asks us to reconcile our beliefs about science and faith.

ONE CAN'T HELP but say that George Church is a towering figure in biology. Standing six-foot-five without his shoes on, he doesn't quite fit through all the doors on the MIT and Harvard campuses, where he heads labs and is a professor. He has a broad, cherubic smile, rosy cheeks, thick white hair, and a long, fluffy beard. Santa Claus's equally affable geneticist brother, basically. Given Church's research, others compare him to Charles Darwin, or to even grander figures. During a sprawling conversation about engineering the future of life using synthetic biology, the comedian Steven Colbert once interrupted Church with an urgent question: "Do we need reinventing?" He went on: "We were invented once. By God, the Almighty Father, maker of heaven and earth. Are you playing God, sir? Because you certainly have the beard for it."[3] Whether Colbert realized it or not, the comparison wasn't entirely a joke, because Church is deeply involved in efforts to create new life forms and resurrect the dead.

Church was born on MacDill Air Force Base in Florida in 1954 and grew up in nondescript middle-class neighborhoods near Tampa Bay. His father was a lieutenant in the air force—as well as a race-car driver, a barefoot water-skier, and someone far more interested in adrenaline-fueled pursuits than a quiet home life. His mother, an attorney, psychologist, and author, was a brilliant thinker who tired of such antics. She remarried twice, the final time to a doctor named Gaylord Church. He legally adopted George, who was then nine. George was instantly captivated by the medical gadgets in his stepfather's bag. Gaylord showed his curious son how to sterilize needles and sometimes let George inject him with actual medications.[4]

At the same time, Church was giving his Catholic school teachers headaches. He was polite, but he had many more questions than the nuns were prepared to answer, often leading his teachers down deep theological rabbit holes. For high school, he went to the Phillips Academy, the venerated boarding school in Massachusetts that Marvin Minsky had attended, and it proved to be a better fit. There, he became immersed in computers, biology, and math—but he increasingly found himself unable to fully fall asleep at night or stay awake during the day, even in the math classes he loved. The other students teased him relentlessly about it. Worse, his algebra teacher finally told him not to bother coming to class anymore: if he was going to fall asleep in class so much, he could learn the material on his own. He felt ashamed that he'd disappointed his teacher, and hated the idea of not blending in with the crowd.

The problem persisted after he enrolled at Duke University. In a meeting, or in a seminar, he'd unintentionally doze for a few minutes. Upon hearing his name, he'd snap awake and respond as though he'd been alert and attentive the entire time. Once, a department head, incensed that a student dare doze off, threw chalk at him. Still, Church completed an undergraduate degree in chemistry and zoology in just two years. Then he stayed for graduate school and studied biochemistry. He soon got sidetracked with crystallography, a new way to study the three-dimensional structure of transfer RNA, which decodes DNA and carries genetic instructions to other parts of cells.[5]

Church continued to have problems with his sleeping and waking cycle. Most people assumed he fell asleep because he was bored or just daydreaming. In truth, he was involuntarily slipping quickly into REM sleep—that phase of sleep where dreaming occurs—and bringing his waking thoughts with him. In these lucid dream states, he had visions of alternate futures, where he explored different permutations of scientific solutions with wild, weird applications of technology no one would think to consider outside of a dream state.

As a student, Church's intellectual curiosity and willingness to allow his mind to wander (not to mention the sleeping thing) often got him in trouble. He spent so much of his time—a hundred hours a week or more—on his groundbreaking crystallography research that he never showed up to his core classes, which, unsurprisingly, he failed. The biochemistry program dropped him, and he attempted to transfer to other departments to pursue his research. But the mishmash of the classes he'd taken, his oddball reputation, and his weird area of research didn't resonate with professors. He was now twenty. He'd published notable research papers, and had been awarded a prestigious National Science Foundation fellowship. But he found himself thwarted by the bureaucracy of academia.[6]

Nevertheless, Church managed to transfer to Harvard, resolving to finish his degree. One early fall day in his first semester, he was running a few minutes late to one of his classes, so he quietly snuck in and took a seat in the last row. He pulled out his notebook and glanced up at the slide presentation that had already begun. What he saw astonished him: the lecture that day centered on a paper Church himself had written. His professor, the pioneering molecular biologist Walter Gilbert, didn't realize Church was in class. (Three years later, Gilbert would win the Nobel Prize for developing one of the first methods for sequencing DNA.)

Church kept dreaming about biochemistry and came up with a bunch of bold ideas. One involved a machine for fast, cheap DNA reading. Another centered on rewriting genomes using off-the-shelf molecules as a way to improve upon nature's creations. He imagined enzymes that could edit parts of genomes, and had visions of neurodiverse people—people with, say, obsessive compulsive disorder and autism—who could be given the ability to turn their special abilities up or down, rather than dampening them with medication. These ideas made their way into the lab, where Church focused on genomic sequencing and on molecular multiplexing, a technique that sequenced multiple DNA strands simultaneously, rather than just one strand at a time, the accepted method at that point. The technique wasn't new, but scientists hadn't pursued it, because it sounded absurd to them. Church proved it could work, and it caught on, drastically reducing the cost of DNA sequencing.[7]

Along the way, he met the molecular biologist Chao-ting Wu, a Harvard PhD. She admired his unbridled work ethic and creativity, and she supported his wild ideas. They fell hard for each other and married in 1990. A few years later, they had a daughter, who also developed unusual sleeping patterns. Wu suggested that they both get evaluated, and both Church and their daughter were diagnosed with narcolepsy. Church realized that the standard treatment might rob him of his lucid dream states, so he decided to live with the symptoms. He stopped driving, but he also learned tricks to stay alert, like standing and shifting his weight between his feet.[8]

Inspired by his family, who'd helped him to flourish despite his idiosyncrasies, Church became a champion of others' ideas. By the early 2000s, he and a sprawling diaspora of acolytes had published hundreds of papers, many of which laid the foundations for present-day synthetic biology. A 2004 paper made the case for cheaper DNA synthesis and showed how to print strands onto microchips.[9] A landmark study in 2009 revealed a new technology that would enable millions of genomic sequences to be analyzed simultaneously.[10] Church then had an idea about how to speed up the process of constructing and assembling genes by putting evolution to work in the lab. Recall that synthesizing artemisinin required roughly $25 million and some 150 person-years of labor—and that task only involved tweaking a few dozen genes, a far cry from synthesizing an organism. Church thought that instead of writing the perfect DNA from scratch, there could be a machine that started with a rough design, developed multiple variants automatically, and then selected the best versions.

He and a small group in his lab went on to create a machine to do just that. It was a hodgepodge system of robotic arms, flasks, tubes, wires, and sensors, with a computer to run it all. Their first experiment involved tweaking a strain of *E. coli* to have it produce more lycopene, the kind of carotenoid that makes tomatoes red. The process resulted in fifteen billion new strains, all with genetic tweaks, and some with five times as much lycopene as the original. Church called this approach multiplex automated genome engineering (MAGE). It was evolution, but supercharged. He imagined practical applications, like creating human cell lines with different variations that could be studied. With this method, scientists could learn how mutations cause disease, for example. It could radically change our approach to medicine. We could engineer stem cells that were resistant to viruses and use them in cell-based therapies. Or we could engineer and grow new organs that would be resistant to disease. Theoretically, we could create virus-resistant babies by tweaking genomes and using IVF to implant embryos.

But, perhaps most notably, in 2012 Church helped lay the foundation for CRISPR, that cornerstone technology for gene editing, by figuring out how to easily alter DNA sequences and modify gene function. "CRISPR" stands for "clustered regularly interspaced short palindromic repeats," which refers to certain types of repeating DNA sequences in a genome that read the same backward and forward. But, more broadly, it is a technique with widespread applications, from correcting genetic defects to creating hardier plants or wiping out pathogens.

Church and his former postdoctoral student Feng Zhang, of Harvard's Broad Institute, published papers in the journal *Science* demonstrating how to use the CRISPR technology to guide a bacterial enzyme, Cas9, to precisely target and cut DNA in human cells. Their papers built on some earlier discoveries by microbiologist Emmanuelle Charpentier and biochemist Jennifer Doudna, who at the time were at the Umeå Centre for Microbial Research in Sweden and the University of California at Berkeley, respectively. Charpentier and Doudna had shown how to use enzymes known as CRISPR-associated proteins to effectively cut and paste DNA.[11] Their system inspired a gold rush in the 2010s and resulted in the Nobel Prize in Chemistry in 2020—it was the first time an all-women team won a science Nobel.[12] Church, whose role did not receive the same adulation, didn't mind the shadow they cast, telling a reporter, "I think it's a terrific choice. . . . [T]hey made the key discovery" and went on to praise Charpentier and Doudna for their work.[13]

For the past two decades, Church has cofounded a new company every year, on average, primarily as a means to get his most promising postdoc fellows out of the lab and into the real world. He's filed sixty patents and mentored a new generation of genetic engineers, whose work is shaping the world of tomorrow.[14] In the mid-2000s, he started thinking about reinventing the plastic cup—without using petrochemicals. Essentially, Church and his team genetically reprogrammed microbes so they would eat sugar and excrete polyhydroxybutyrate, a strong and biodegradable material that could hold liquids for a short period of time. They were the perfect candidates for concession stands, in other words—and in 2009, they made their debut at the Kennedy Center during intermission, emblazoned with a label that proudly declared "Plastic made 100% from plants."[15]

Church was also part of a small group of scientists who proposed the idea of the BRAIN Initiative, an ambitious public-private collaborative effort between the National Science Foundation, DARPA, and other institutions to understand how the brain works. In 2005, he launched the Personal Genome Project, a clearinghouse for public genome, health, and trait data.[16] As part of this project, he and other prominent voices in the scientific community—including investor/philanthropist-trained cosmonaut Esther Dyson; Harvard Medical School's dean of technology John Halamka; Rosalynn Gill, the founder of the personalized healthcare company Sciona; the noted psychologist and writer Steven Pinker, and, of course, Church himself—made data from their own genomes public, in a bid to foster accessible research into the genes and traits that shape us, and to spur a dialogue about transparency versus privacy when it comes to our individual genetic codes.[17,18,19,20] Consider the implications of such prominent people making their genetic data accessible: ten genomes isn't exactly a big dataset, and although the data were anonymized, their identities were publicly announced. There was no real way to guarantee absolute privacy. They volunteered because it was Church who asked.

## REANIMATION

By now, you've likely gathered that Church is a brilliant and provocative thinker, an inspiring mentor, and, perhaps, prone to taking on too many projects for any one human to handle. And that he'd be just the kind of guy, in other words, who'd figure out how to resurrect extinct animals—specifically the woolly mammoth, which disappeared around four thousand years ago, during the Pleistocene era.

Before then, woolly mammoths roamed the northernmost parts of the planet for millennia. Picture a close cousin to the elephant, but with coarse fur and layers of fat to protect them from the cold of the ice age, and with long tusks that helped them forage for food. (Much later, they inspired the fictional banthas in *Star Wars*.) Exactly why the mammoths went extinct is unclear, but researchers point to a combination of hunting and temperature fluctuations, which both depleted the herds and reduced their food supply.

Woolly mammoths were a "keystone species," one that other species in the ecosystem depended on in many ways for stability. They stomped around in herds, knocking down trees and packing down snow layers as they searched for dead grasses to eat, and that helped keep the permafrost layer stable. Once the mammoths, and other large grazing animals, stopped compacting the snow and eating dead grasses, the ecosystem began to change: the snow melted more easily, which allowed the sun to reach the permafrost. The permafrost layer is now melting at an alarming rate and releasing greenhouse gases into the atmosphere, which creates a vicious cycle: hotter temperatures lead to more melting, which releases more gases, which causes hotter temperatures, and on and on it goes. Raising the woolly mammoth from the dead and rewilding it in Canada and Russia could restore the ecosystem, and—let's face it—be a novel and ultra-cool defense against the existential threats posed by climate change.

Church has thought a lot about how de-extinction might work. He wasn't the first to take a stab at it, though. In 1996, Dolly the sheep, the world's first cloned mammal, was born.[21] Thanks to a technique called nuclear transfer, her birth helped open up the prospect of bringing back extinct creatures. In this technique, the nucleus of an intact cell is carefully extracted and inserted into the egg of the same or a closely related species. The rest is similar to an IVF procedure: the hybrid egg is implanted into the uterus for gestation, and, if things go well, the result is a healthy, live birth. In 2000, the last surviving Pyrenean ibex, a type of wild mountain goat, died. But cells from that last animal were frozen in liquid nitrogen, and, thanks to the nuclear transfer method, in 2003 researchers successfully cloned a calf and brought it to life—albeit just for a few minutes.[22] The method has the possibility of working only if fully intact, functional genomes are available—from, say, an unusually well-preserved frozen carcass. And it just so happens that excellent specimens of woolly mammoths dot the Arctic Circle. Even then, it's far from certain that an effort to revive an extinct species would be successful. A clone might not survive. An animal that went extinct thousands of years ago likely won't have a genome adapted for living on Earth today.

That's why Church envisioned a different approach: starting with a fully intact healthy cell from a closely related species and working backward, with genetic fragments from preserved specimens.[23] Take the passenger pigeon, for example, which has been extinct since 1914, even though it was once so prevalent in the United States that millions would block out the sun as they flew overhead.[24] The passenger pigeon could be brought back using the stem cells of a current close relative, the rock pigeon. Which means that some of the genes from the passenger pigeon could be inserted into stem cells, transformed into sperm cells, and then injected into eggs to develop into embryos. The result would be a common pigeon with some traits of the passenger pigeon.

Such notions fascinated Stewart Brand, the legend in tech who created the *Whole Earth Catalog* and the pioneering online service The WELL, and the biotechnology executive Ryan Phelan (who is also Brand's wife). Brand, Phelan, and Church collaborated to launch an effort to de-extinct keystone animals, including the passenger pigeon and the woolly mammoth. Or, to be more precise, a *mostly* woolly mammoth, since bringing it back would involve splicing the genes of the extinct species into the stem cells of a modern Asian elephant, the mammoth's closest living relative.

The idea of de-extinction gained currency at a special TEDx De-Extinction conference in 2013, which brought together molecular biologists, conservationists, and journalists to discuss the prospects of bringing back woolly mammoths, Tasmanian tigers, and other species. There, Brand gave a provocative talk about the loss of biodiversity and the promise of reanimating extinct animals using Church's technology. He used the conference and the TED platform to launch Revive and Restore, an initiative to investigate the causes of extinctions, to preserve biological and genetic diversity, and to repair our ecosystems using biotechnologies.[25]

Brand's TED talk was wildly popular.[26] It also sparked outrage among many, including scientists, conservationists, and others who were horrified at the prospect of bringing back long-extinct creatures. This wasn't cloning—or merely making a copy of something that used to be alive. Rather, it was blurring the clear distinction between extant and extinct. And Church also made clear that he wasn't just interested in mammoths and pigeons. He also wanted to tinker with the DNA of Neanderthals—not to resurrect a different species, but to improve our own.[27]

You might believe, as scientists once did, that Neanderthals were primitive subhumans, cruder, brutish versions of ourselves. But more recent study has revealed that the Neanderthals were highly intelligent. They had built a well-organized civilization, and, as species go, were quite successful: they lasted 250,000 years. (The oldest *Homo sapiens* are believed to be 300,000 years old.) Neanderthals' bodies conserved heat efficiently, which meant they could survive hostile environments. They were incredibly strong—that part of the stereotype is true—but they also had excellent fine motor skills. Creating a newer Neanderthal, by crossing *Homo sapiens* with *Homo neanderthalensis*, could result in a heartier species of people who could better withstand the modern challenges of climate change and extreme weather events and would be more likely to survive a necessary migration to a new and vastly different environment.

Several Neanderthal genomes have been sequenced from DNA in fossils found across Europe and Asia. Analyzing and synthesizing the genome in smaller bits would eventually enable scientists to assemble the correct sequence in a human stem cell, which, theoretically, would result in a Neanderthal clone. But let's let Church explain it:

> You'd start with a stem cell genome from a human adult and gradually reverse-engineer it into the Neanderthal genome or a reasonably close equivalent. These stem cells can produce tissues and organs. If society becomes comfortable with cloning and sees value in true human diversity, then the whole Neanderthal creature itself could be cloned.[28]

There would be challenges, of course, for the modern Neanderthal. To name one, the typical Western diet is heavy on dairy, refined grain products, and heavily processed foods. Taco Bell's Nacho Cheese Doritos Locos Taco—the uninitiated should envision a taco shell made out of Doritos filled with seasoned cheap meat and a cheddar cheese amalgam made with an anti-cracking agent—is rough on even the most battle-tested stomachs. As tough as Neanderthals are, a couple of those would likely lay them, and their prehistoric GI tracts, flat.

Perhaps resurrecting the Neanderthal sounds like a bad idea. But what if we could borrow just a few Neanderthal genes to tweak our biologies just a bit? Neanderthals didn't have celiac disease, for example, a painful condition for modern humans who are allergic to gluten. Their immune systems react differently than ours, which could offer researchers insights into how to cure autoimmune diseases like rheumatoid arthritis, multiple sclerosis, and Crohn's disease. And Neanderthals' bones were extremely strong, so we might borrow those genes for bone density to help treat osteoporosis, which affects hundreds of millions of women as they age.

IF MIXING NEANDERTHAL genes with ours and implanting them in a gestational surrogate sounds like a horror movie or a work of dystopian science fiction, you're not far off. Most often, when humans intervene in God's grand plan, disaster follows: H. G. Wells's *The Island of Dr. Moreau* (1896), Aldous Huxley's *Brave New World* (1931), Frank Herbert's *Dune* (1965), Ursula Le Guin's *The Left Hand of Darkness* (1969), Nancy Kress's *Beggars in Spain* (1991), Richard Morgan's *Altered Carbon* (2002). It's been a persistent and recurring theme in *Star Trek* and in Marvel's X-Men series, in which the villain Magneto hatches a plan to "make *Homo sapiens* bow to *Homo superior*!"

Historically, neither science nor society likes it when people play God, or even play with the issues around playing God. Mary Shelley had created a story about a monster—not an actual monster, mind you—and her story was so subversive that she feared the government would take her children away from her if her name was attached to the work. When Dolly the sheep was successfully cloned, the event triggered emergency meetings and press conferences worldwide. Almost everyone missed the stated point of the Dolly project, which was to advance our understanding of how cells changed during their development. Instead, the reaction was swift, and wildly negative. Dr. Ronald Munson, a medical ethicist at the University of Missouri in St. Louis, told the *New York Times*, "The genie is out of the bottle." What was next, he demanded—cloning Jesus Christ using a drop of blood from the cross?[29] Professor George Annas, the chair of the Health Law Department at Boston University's School of Public Health, admonished the biology and genetics community. "The reaction should be horror," he said, arguing that full-blown human cloning was the next logical step. "Parents do not have the right to collect cells from a child to reproduce that child. The basic public outcry against human cloning is correct."[30] The Church of Scotland issued an official decree demanding that the United Nations pass an enforceable ban on cloning. Quoting Jeremiah 1:4–5 in the Old Testament, the church argued that mankind could not take the place of God: "The Lord . . . [said] . . . 'Before I formed you in the womb I knew you, before you were born I set you apart.'"[31] Then president Bill Clinton held a televised event to announce a new ban on federal funds for any research project involving human cloning.[32]

A CNN/Time poll released on March 1, 1997, revealed that the majority of Americans suddenly had decisive opinions on nuclear transfer technology, a form of cloning.[33] It's difficult to imagine that, before Dolly, they had ever thought much about cloning or that technique at all. A third of the respondents said they were so disturbed by Dolly's mere existence that they would join public demonstrations and protests. Meanwhile, in the nearly quarter-century since Dolly, we've gained important knowledge, new biological technologies, and a broader understanding of how life works. The last time we checked, the planet has not been overrun by demon sheep. Because of Dolly, scientists began cloning adult stem cells, which led to the creation of artificial induced pluripotent stem cells (iPSCs) that could be used in medical research. That reduced the need for embryos, which have long caused ethical concerns for many people. Researchers were able to study the aging process: this was the first time an adult cell had been reprogrammed and made to act as though it was youthful again. This work opened the door to all sorts of new stem-cell-based therapies in humans: if a remedy was derived from one's own genetic code, there would be no chance of the immune system rejecting it. Today, there are several regenerative therapy treatments available to treat blood-related diseases, including leukemia, lymphoma, and multiple myeloma, as well as other degenerative conditions, such as heart failure.

Our beliefs and perceptions can be slow to change, which is understandable—centuries of writings and deeply held social values influence how we think. There is often little warning ahead of groundbreaking scientific announcements, so we may feel shocked, confused, or even anxious with news that challenges our mental models. Sometimes that even happens within the scientific community itself. As Church's views on de-extinction became more widely known, the editorial board of *Scientific American* wrote a scathing rebuke in 2013, which centered mostly on how the money spent on this experimental technology would be better spent on traditional conservation efforts.[34] Church responded with his own *Scientific American* article, in which he calmly explained that the goal of reanimation, and his project, wasn't to make "perfect living copies of extinct organisms, nor is it meant to be a one-off stunt in a laboratory or zoo." The point, he explained, was to learn how we can adapt existing ecosystems to the environmental changes we've caused to ensure our own survival.[35]

As of December 2020, Church and his team of Harvard researchers were making progress toward their mammoth (sorry) goal. While the Asian elephant's genome is roughly 99.96 percent similar to the woolly mammoth's, that 0.04 percent difference adds up to 1.4 million differences in the DNA. Most aren't consequential, but as of this writing, Church's team has identified 1,642 genes that are. That work continues: currently, they're also painstakingly designing, testing, and tweaking those lab-grown cells one at a time to produce the right genetic sequence for a mammoth-like Asian elephant to survive. They're aiming for a common elephant base, but with the mammoth's thick hair; its hemoglobin, which is uniquely adapted for cold weather; its ability to store layers of fat; and other improvements, like cells that can pass sodium ions through their membranes, making them better adapted to harsh winter environments.[36] Once they have tweaked the right combinations of characteristics, they can begin the work of injecting these skin cells into stem cells, and producing living, mostly woolly mammoths. In September 2021, Church and Texas entrepreneur Ben Lamm launched Colossal, a business venture to support the woolly mammoth project.

And if they are successfully created, these twenty-first-century woolly mammoths will have a home in a place inspired by novelist Michael Crichton, though named after the Pleistocene period rather than the Jurassic. Pleistocene Park (really, that's the name) is an experiment in Siberia, a nature reserve that has been rewilded with Yakutian horses, elk, bison, yaks, and other native species after years of industrialization had decimated those populations.[37] Modified mammoths could demonstrate whether heavy beasts stomping down on the snow and permafrost could make a dent in our climate woes.

It may seem that reanimating and rewilding these beasts would put humans more decisively in the role of the Creator than ever before. In reality, we've been playing the role of Creator for millennia. The real problem is that we've been doing a terrible job of it.

AS EUROPEANS SAILED the Atlantic in the fifteenth century, they systematically altered the environment. They not only discovered "new worlds" (or, more precisely, worlds new to Europeans), they also introduced their own indigenous plants, animals, and diseases to the people they encountered. In 1492, Christopher Columbus landed in present-day Dominican Republic. A year later, he returned with 1,500 people, dozens of seeds and cuttings of plants (wheat, barley, onions, cucumber, melons, olives, and vines), and hundreds of horses, steer, and pigs. This began what's known as the Columbian Exchange, a great reshuffling of organisms traversing the oceans to new lands. As strange as it sounds today, before the Columbian Exchange, the capsicum peppers native to Brazil weren't yet staples in spicy Indian food, and no one in Ireland had yet tasted a potato. Before the sixteenth century, there were no peanuts in Africa, and no tomatoes in Italy. There was no wheat in North America before the Columbian Exchange. It radically transformed agriculture, changed diets (of people and animals alike), and reshaped lives and cultures.[38]

But as Europeans arrived in unfamiliar lands, they also brought with them dangerous pathogens to which the indigenous peoples had no natural immunity. Each new ship of people also brought new waves of infectious diseases: smallpox, pneumonia, scarlet fever, malaria, yellow fever, measles, whooping cough, typhus, and rhinoviruses. The new diseases killed 80 percent of all indigenous peoples and decimated great swaths of local animal and plant populations. (Columbus and many of his compatriots also fell ill.)[39]

That exchange eventually gave rise to a global economic framework. The dystopian next-order consequences that followed now bedevil humanity in ways that the seafarers from hundreds of years ago could never have imagined. The spread of disease, conversion of land for industrial-scale agriculture, overhunting of animals, centuries of mining, and pollution generated by our global trade routes laid the foundation for the biodiversity losses and climate change we are experiencing today.

There have been 7,348 big natural disasters around the world from 2000 to 2019, many caused by climate change.[40] The number of major floods doubled. Storms severe enough to cause significant property damage increased 40 percent. The historic 2019–2020 wildfires in Australia blew so much ash and sediment into the atmosphere that it blocked sunlight and caused a sort of nano ice age: weather researchers reported that the Earth temporarily cooled a fraction of a degree (the exact amount is not yet known). Pyrocumulonimbus clouds—fire-filled thunderclouds—spun up their own winds, which resulted in dangerous vortexes in other countries. Wildfires have thus far resulted in almost $3 trillion in cumulative global economic losses.[41]

It is estimated that the world's cities will warm as much as 4.4°C (roughly 8°F) by the year 2100. That's a tremendous amount of additional energy in our weather systems. It may not seem catastrophic to you if you live in a home with good climate control and good air circulation. But in cities like Paris and London, with metro areas that are home to 25 million people, many old buildings don't have cooling ducts and can't support air conditioning units. A few hot summer days cause serious problems for human life. When you start to overheat, your body redirects the blood usually flowing to your organs back out to your skin in an effort to cool you down. In sustained, extreme heat, that process doesn't stop, and a lack of blood flow to your organs causes them to shut down. If it's both hot and humid, the biological process your body would normally undertake to cool you—sweating—no longer works. Scientists at the University of Hawai'i at Manoa have categorized twenty-seven distinct ways that a heat wave can kill even the healthiest people.[42] Our current climate trajectory means that the majority of people alive in 2100 will be regularly exposed to heat that can kill.

Consider, too, that in the next few decades the world's population is expected to increase from 7.7 billion to 9.7 billion. Even if we didn't have to contend with the agricultural impacts of climate change, we're not in a position to feed another 2 billion people. Much of that population growth is projected for India, where roughly 10 percent of the food grown is irrigated using rapidly dwindling nonrenewable groundwater sources. The same is true in California's Central Valley and northeastern China and Pakistan. These are major food-producing areas responsible for cereal grains, vegetables and fruit, cotton, hay, and rice—staples that we, and our economies, depend on. For all the advances in agriculture that we've wrought, roughly 80 percent of the world's crops depend on rainwater, so farming still requires predictable weather patterns. Extreme precipitation events wreak havoc on the world's food supply. If one accepts evidence-based research around climate change, population growth, and extreme weather events, the logical conclusion is that our current path will lead to a lot of unnecessary death, at best, and worldwide famine and chaos, at worst.

In addition, our planet's biodiversity has plummeted. The total biomass of the human race accounts for less than 0.01 percent of all life on Earth—that is, we make up less than one one-hundredth of living organisms—but we've wiped out 83 percent of the animal species. The UN's Intergovernmental Science-Policy Platform on Biodiversity and Ecosystem Services published an apocalyptic dataset in 2019 concluding that 1 million animal and plant species were at risk of destruction. Not 1 million rabbits, or 1 million daffodils, but 1 million *species*. Gone.[43]

This is the world that we created, not through intelligent planning and intentional design, but through a mindless evolution made through centuries of choices and actions, all of which continue to play out in unpredictable ways. To cite one small example, consider how rising sea levels have changed the landscape and fauna of the southeastern United States. They created the perfect conditions for a tiny, inconspicuous purple burrowing crab (*Sesarma reticulatrum*) to thrive. Its population exploded and started feasting on cordgrass, a hearty native plant that does much to keep coastal marshland in place. Now, what had been contiguous grasslands are breaking apart and enlarging tidal creeks.[44] There's more sediment in local water streams, more flooding during storms, a loss of recreational space for water sports, and fewer commercial fishing grounds. These crabs have done so much damage already that you can see it from space. They've also forced scientists to reorganize the hierarchy of their ecosystem: *Sesarma* crabs are now considered a keystone species.

It's time—in fact, it's past time—to start investigating alternate stories that contradict our prior beliefs, that force us to evaluate risk objectively, rather than emotionally. Here are just two examples showing how synthetic biology can potentially contribute:

**Upgrading humans:** We already do this, but we don't refer to it as "upgrading." Humans are born with vulnerabilities to certain viruses, including rotavirus, hepatitis A, hepatitis B, polio, pneumonia, *Haemophilus influenzae* type b, varicella, measles, mumps and rubella, diphtheria, tetanus, and pertussis. In the United States and other developed nations, most babies receive "upgrades"—in the form of vaccines—within their first year. Adults get an annual flu shot; at fifty they may get the shingles vaccine. In early 2021, many of us were clamoring for the most important upgrade in our lifetimes: the COVID-19 vaccine.

What could come next? We could take a cue from Church and develop tools that allow neurodiverse people to enhance their lives. When Amy was in her thirties, she was diagnosed with obsessive-compulsive disorder (OCD), a clinical diagnosis that confirmed what she'd long suspected. She would go through crippling phases where she would silently count her steps, or find herself unable to stop reworking statistical models, or feel compelled to take the exact same route to work without deviation. These loops sometimes forced her to isolate away from other people, but they also gave her an almost superhuman level of stamina: she could sustain her focus and work on hard problems for days at a time without needing a break. OCD is the result of serotonin mucking up the communication lines between the front part of the brain and deeper structures located in the interior. Treatment involves medications that normalize serotonin, or cognitive behavioral therapy, a type of talk therapy that rewires neurotransmitters over time, or both. A serotonin inhibitor would have stopped Amy from looping, but it also would have dampened her creativity and drive. There would be no rheostat—no way of dialing up or down her serotonin as needed. With the advice of her doctors, Amy chose cognitive behavioral therapy, but she thinks often about Endy and Knight's repressilator—perhaps it could enable her to make use of OCD as needed. It would become a feature, rather than a bug, in her system.

**Upgrading agriculture:** Almonds are a great food—tasty, nutritious, usable in many forms. But they make for a notoriously thirsty crop. The almond tree's genome was sequenced in 2019; it could be tweaked to require less water, produce twice as many nuts per plant, and grow as a compact plant rather than as a big, bushy tree.[45]

Then it could be cultivated in an entirely different way. Almost all of our food is grown outside, where we have no control over the weather. Genetic editing, custom microbes, and precision agricultural systems that make use of artificial intelligence and robotics are allowing us to explore alternative means of production—such as vertical farms, in which crops are grown in stacked layers inside of enormous warehouses. LED lights act as the sun, while sensors and artificial intelligence systems monitor hydration and nutrients, adjusting levels as needed. Robots move along rows of snow peas, garlic, spinach, and lettuce. Such operations tend to deliver ten to twenty times the total yield of conventional farms with far less waste.

WE CAN INTELLIGENTLY design the next phase of our evolution—in fact, soon we may have no choice—but only if we are willing to reconsider our long-standing ideas about creation and creator. Today, some religious scholars and faith communities believe that synthetic biology (as well as other sciences and technologies) are examples of growth and human progress, and that God's mission involves the transformation and renewal of creation. Working against illness, hunger, and death are important values across numerous religions. Synthetic biology is a natural expression of that progress.

Dr. Francis Collins, director of the National Institutes of Health, is a committed Christian. He has written best-selling books on the relationship between science and religion, including *The Language of God*, in which he makes a scientific argument for the existence of God, and *Belief: Readings on the Reason for Faith*, in which he examines the mysteries of God and faith. Like George Church and Craig Venter, Collins is a pioneer in genetics who has contributed significant research to the field. His appointment by President Barack Obama as NIH director in 2009 put him in charge of 20,000 scientists and staff, 325,000 outside researchers, and 27 institutes and research centers.[46] While there, he permitted, and advocated for, stem cell research, despite the objections of many who share his faith—indeed, despite his own personal faith-based concerns. Collins, who has served under Presidents Obama, Trump, and Biden, continues to make the case for editing genes.

If we, like Collins, approach alternative futures objectively rather than emotionally, we can make space for both God and genome alterations. If we, like Church, allow our minds to wander productively, a new world awaits. One marked by a new bioeconomy, innovative scientific solutions to thorny problems, and countless clever ways in which we can categorically improve—even save—life as we know it.

# PART TWO | Now

| FIVE |

# THE BIOECONOMY

IN LATE 2019, AS THE CHINESE GOVERNMENT TOOK ACTIVE MEASURES to downplay evidence of a newly emerging and extremely dangerous novel coronavirus, Dr. Zhang Yongzhen, a virologist at the Shanghai Public Health Center, was growing increasingly concerned about the mysterious pathogen. He and his team had already discovered more than two thousand viruses, but this new virus ravaging the city of Wuhan, just eight hours due west of his lab, had troubling characteristics. Among many other things, someone could be infected by it for ten to fourteen days before they developed symptoms. If the government didn't take action fast, Zhang feared, it would spread quickly from the people in Wuhan—Chinese citizens, foreign workers, and tourists—to the communities where they traveled. Like researchers at many other Chinese labs who knew about this new pathogen, Zhang and his team started working to sequence the coronavirus.[1]

Zhang's team, relying on its deep experience discovering new viruses and modern technology, decoded the genome in just forty hours.[2,3] As they suspected it would, this coronavirus resembled the severe acute respiratory syndrome virus (SARS) that ravaged many countries in 2003.

As the Chinese government delayed acknowledging the dangers of this coronavirus, Zhang weighed the personal and political consequences of publishing the sequence against the shrinking opportunity to stop the virus from spreading. Luckily for the world, he didn't wait long. On January 5, 2020, Zhang published the sequence in GenBank, which is essentially a biological version of Wikipedia.[4,5] There, researchers submit genetic sequences they've decoded along with notes about them. Community moderators review the sequences, and if they're approved, publish them. (As with Wikipedia, nobody gets paid for the work, but everyone benefits.) But Zhang was concerned that even with that process, which required far less time than traditional peer-reviewed journals, it would still take too long to get the word out. An Australian friend reached out and asked to publish a message on Virological.org, a far more free-wheeling discussion site. Think Reddit, but for virologists. On January 10, 2020, that post went live, immediately catalyzing swift action on GenBank.[6]

Days later, the entrepreneur Noubar Afeyan was out for dinner in Cambridge, Massachusetts, celebrating his daughter's birthday when he received an urgent text message from Stéphane Bancel, his CEO. Afeyan had founded Moderna, which was then still pretty obscure. Its name combines the words "modified" and "RNA"—referring to the idea that messenger RNA could be re-engineered using synthetic biology techniques to develop personalized cancer treatments. The technology worked in a lab, but it hadn't yet led to any marketable products, so his team had shifted to engineering mRNA for other treatments, such as developing new bits of code to teach our cells to perform new functions, including developing antibodies and healing tissue.

Afeyan rushed out into the cold to call Bancel back. Bancel had been following the discussion about Zhang's sequence and had been working in the background to design an mRNA vaccine for the new virus. With designs ready to go, Bancel needed permission to divert resources from the twenty products Moderna had in development into an mRNA vaccine for this new virus. They had ample experience: Moderna had been working with the NIH to prototype mRNA vaccines for coronaviruses. But they hadn't yet brought any such vaccines to market, and Moderna didn't exactly have piles of extra cash on hand to crank up a full-scale vaccine program from scratch. Despite these risks, Bancel had sensed something big was happening, and Afeyan agreed. "Just get moving," Afeyan told him.[7]

Working with the sequences that Zhang had published on GenBank, Moderna began designing specific RNA sequences for the coronavirus spike protein. As you know all too well, the virus resembles a ball covered in spikes. Based on the team's prior research, they knew that those spike proteins would be most likely to trigger a response: the human immune system tends to fixate on big, obvious cues. In essence, Moderna's vaccine would act like a biological most-wanted poster from the Old West that read, "Wanted! Keep an eye out for this ball covered in spikes! If found, kill on sight!"

Moderna had already done much of the hard work to develop mRNA vaccines, like figuring out how to direct that mRNA to the outer region of cells, the cytoplasm, which is where proteins are constructed. The mRNA would set up a temporary camp outside the cell's nucleus, translate its own string of letters into a protein, and leave shortly after. This would spur the cells to produce the safe components of the coronavirus in order to kick-start the immune system. Then the mRNA would degrade, leaving the body to fight on using its own defenses.

Using custom-built tools, they looked for certain features in the genetic code: on and off switches, nucleic acid sequences that mark the end of a gene during transcription, and sets of instructions that define the start and stop points for various proteins. It took only two days to crack the code. The difference between SARS-CoV-2 and other coronaviruses is just twelve extra letters in its genome: CCU CGG CGG GCA. Those twelve letters are what make it so virulent. They're what allow the spike protein to be activated and to invade human cells. But mRNA could deliver a set of instructions to cells to target that string of letters and thwart the virus's attack.[8]

This approach—using synthetic RNA—would be far more effective and adaptable than long-standing vaccine protocols, like making use of weakened viruses, or, as with each year's flu vaccine, needing millions of eggs to produce the necessary doses. In effect, Moderna was crafting genetic instructions that could be written like software and packaged into the equivalents of nanoscopic USB drives. Once these biological USB drives were inserted into the cells, those cells would dutifully download the mRNA instructions and follow them. Such vaccines would also be safer and easier to control. Unlike gene therapies, which can lead to permanent or even inherited genetic changes, mRNA only exists in our cells ephemerally. The vaccine-generating program would run for a short time and self-destruct.

Moderna, along with the biotech startup BioNTech, finally found a tangible application for what everyone thought was impractical, fantastical technology. But the vaccines we have today wouldn't have been possible without decades of advances in the machinery and technologies that underpin the processes of synthetic biology. All those technologies, machines, ancillary systems, and what they produce make up the bioeconomy: the production and consumption activities related to and derived from synthetic biology, which collectively fulfill the needs of those companies operating within it. The mRNA vaccines for COVID-19 are just the first of many wonders that tomorrow's bioeconomy will create.

Does that last sentence sound fanciful? History tells us it isn't.

ALEXANDER GRAHAM BELL and Thomas Watson stood on stage at Chickering Hall in New York City, showing off a curious wood and metal contraption to a crowd of three hundred spectators. It was May 17, 1877. The two had spent years toiling in obscurity, single-mindedly pursuing what was then a very out-there notion: how to transmit a human voice using the electrical pulses of a telegraph wire. In time, they invented a receiver and a membrane that could turn speech into electrical pulses, then transform those pulses back into speech. Chickering Hall was their first live demonstration of this new technology. They told the crowd that a wire connected their invention to a device in New Brunswick, New Jersey, and then to another device in New York, where a man with a deep baritone voice would sing the popular hymn "Hold the Fort."[9]

Bell spoke into their invention—and the jerry-rigged connections that supported it—and suddenly the audience heard an unseen singer. Some in the audience were convinced this "speaking telephone" was no more than an elaborate trick, and noisily demanded to search backstage for a hidden vocalist. Bell persisted with his demonstration, launching into an exhaustive lecture detailing the electrical design and science of sound. That, and a backstage tour, eventually convinced the doubters that the telephone was real.[10]

It would take a few years for entrepreneurs to realize the value of this groundbreaking technology, but in time dozens of businesses were launched to support the demands of the new telephone economy: common battery systems, metallic circuits, cables, switches, handsets and wall-mounted telephones, network designs for telephone exchanges, enormous antennas to carry telephonic signals and companies to install them, switchboards and operators, utilities to generate and distribute power, even specialized artisan firms that decorated the original telephone exchanges, which were scaffolding that connected wires to grounded poles, with ornate garlands and flags.

By 1918, there were 10 million Bell System telephones throughout the United States and similar local networks spanning Europe and Scandinavia. The problem was that the networks were geographically restricted. The world still needed future inventions, like electronic switching systems, microwave radio technology, and transistors, before telephone calls could cross oceans.[11] While network operators focused on incremental improvements for the next few decades, a member of the United Kingdom's Royal Air Force, named Arthur C. Clarke, had much bolder ideas for communications. In 1945 he wrote a paper describing how to send messages instantaneously between any two points on the planet by using "extra-terrestrial relays" and futuristic telephone exchanges. They wouldn't be tethered to the ground and festooned with pretty garlands, but rather, orbiting Earth, far beyond the horizon.[12] Clarke, who later became the legendary science fiction writer, envisioned an object moving at the same speed as Earth's rotation, positioned 22,236 miles overhead. An antenna on the ground could be pointed at it continuously to provide radio coverage anywhere on the planet.

There was little enthusiasm for such an outlandish idea. It would take a significant world event—the USSR launching a beach ball–sized hunk of metal, called Sputnik I, into orbit—to galvanize support for the project. In 1958, the United States launched Project SCORE (Signal Communication by Orbiting Relay Equipment), a satellite that broadcast a single message:

the voice of President Dwight D. Eisenhower saying, "Peace on Earth and good will toward men everywhere."[13,14] In the 1960s, Bell Labs, part of the American Telephone and Telegraph Company (AT&T), worked with NASA to improve that technology, which eventually led to one of the most important inventions in modern history: a two-way communications satellite that could move in synchronous orbit in relation to a fixed point on the Earth's surface. Such satellites went on to enable broadcast television, accelerate the world's missions to space, provide access to GPS, and make our smartphones possible. Bell's "speaking telephone" invention, once derided as far too outlandish to ever find everyday use, became a communications industry goliath, one worth $1.7 trillion today.[15]

The telephone, and the networks that support it and grew from it, played a defining role in the evolution of nearly every business and facet of society, including another advance that came out of a far-fetched research paper published in 1962. MIT scientist J. C. R. Licklider, who also worked for the Department of Defense's Advanced Research Projects Agency (ARPA), detailed a hypothesis so implausible some early readers thought it was a joke. Maybe it was the name. Licklider proposed a "galactic network" of computers that could talk to one another.[16] His paper came at the height of cold war tensions, and Licklider explained that his network concept could allow government agencies to communicate even if the USSR destroyed the telephone system. (After Sputnik, fears were running rampant concerning what the USSR's technical strengths could inflict upon the United States.) So in 1969, the Defense Department built a pilot network called ARPANET, which linked computers at UCLA and Stanford, to test Licklider's idea. A professor at UCLA tried to send a simple word, "login," to their colleagues in Northern California, but the system crashed after the letter "o." Thus, the first transmission sent via the internet was, disappointingly, the forlorn two-keystroke message "lo."[17]

But by the end of the 1970s, the computer scientist Vinton Cerf had invented the Transmission Control Protocol, or TCP, which acts as a sort of virtual handshake between distant computers and enables them to send information back and forth.[18] That advance paved the way for Tim Berners-Lee's proposal, in 1991, for a decentralized "world wide web," of information where anyone on the planet could share and retrieve information.[19] In 1992, a group of students at the University of Illinois created the Mosaic browser, a friendlier way to search the nascent Web.[20] With Mosaic, users were not limited to a simple text interface, and they didn't need to know how to program a computer. They could see images and use clickable links to navigate to other web pages. The commercial internet followed, as did countless businesses: web hosts; email providers, new networks—such as CompuServe and America Online—along with Google and other search engines and e-commerce sites, including Amazon. The vitriol-fueled social media sites most people think of when we reference "the internet" are a minuscule segment of what's actually there. The internet is the invisible infrastructure upon which we have built nearly every aspect of modern life, from our payroll systems to our city services, health-care records, school tests, and grocery store supply chains.

Several leading economists, including Austan Goolsbee, Peter Klenow, and Erik Brynjolfsson, have attempted to calculate the value created by the internet.[21] They have all come to similar conclusions: any study would be flawed, because the internet is now a general-purpose technology, like electricity. If you remove electricity from society, there would be dire economic consequences for productivity, income, and our ability to create goods and services. The same is true for the internet.

Synthetic biology is now in its speaking-telephone-at-Chickering-Hall stage. It exists. It works. But the vast networks of supporting and related businesses, companies, and ancillary players have yet to be founded. We will someday regard synthetic biology as a general-purpose technology. As with the phone and the internet, the value synthetic biology will deliver to society will extend far beyond what we can conceive of today. But we have some early indicators: the National Academies of Science, Engineering, and Medicine published an exhaustive study in January 2020 valuing the US bioeconomy at 5 percent of gross domestic product (GDP), more than $950 billion.[22] (Note that that was before COVID-19, and the hugely consequential synthetic-biology-driven advances of companies like Moderna.) A May 2020 McKinsey study analyzed the impact on the global economy of four hundred existing synthetic-biology-related innovations in the pipeline today and determined that those advances could yield an average of $4 trillion between now and 2040—each year.[23] And that $4 trillion doesn't account for all the knock-on financial impacts of the adjacent businesses, services, and products that will inevitably emerge to support the industry.

The development of those knock-on impacts is what's known as a value network, an idea that Harvard professor Clayton Christensen introduced in his groundbreaking 1997 book *The Innovator's Dilemma*. Broadly speaking, a value network is a healthy ecosystem of companies working together to create products and services that are attractive to customers. If we took just a small slice of the internet—the Internet of Things—the value network includes software, platforms, interfaces, connectivity, security, agriculture, health care, vehicles, supply chains, robotics, industrial wearables, and dozens of subcategories of businesses, within which there are hundreds of startups and established companies all working to make connected devices communicate better with each other in order to make life easier and more enjoyable for everyone.

Synthetic biology's value network is just starting to form. Even though the sector's response to COVID-19 accelerated growth, so far there are just a few players in each segment. But that's changing quickly. Investors poured $8 billion into synthetic biology startups in 2020.[24] That amount of funding isn't very impressive when you consider that Bytedance, the maker of TikTok, was approaching a valuation of $400 billion in early 2021, but the rate of investment growth in synthetic biology has doubled every year since 2018.[25] There are now exchange-traded funds (ETFs) made entirely of biotech stocks. None existed a few years ago. The investment management behemoth BlackRock launched an ETF for synthetic biology in October 2020, following ARK Capital Management and Franklin Templeton, whose ETFs were performing better than expected. ARK's made a 44 percent return on investment in 2019, and a staggering 210 percent return in 2020. The median size of synthetic biology IPO deals is rising, too: in 2020, the average IPO was valued twice as much as it was in 2019.[26]

All of which is keeping existing companies that support the infrastructure of the bioeconomy busy: hardware manufacturers that make synthesizing machines, robots, and assemblers; wetware companies, which sell the DNA, enzymes, proteins, and cells; and software companies that make special tools, like Photoshop, but for biology.

All these companies require a fast internet connection, automation, and the cloud, not to mention encrypted networks, robust IT services, and database management. Meanwhile, many of the old tools of biotechnology simply aren't good enough to do the kind of high-precision engineering that today's synthetic biology work demands. But, as we're about to explain, the technology and tools that are forming the bioeconomy are evolving rapidly, too.

GIVEN HOW QUICKLY Zhang and his team sequenced SARS-CoV-2, it can be difficult to visualize the old way of sequencing DNA. Before computerized and automated tools were in wide use, DNA had to be carefully prepared, often using homegrown protocols developed by individual labs. The easy-to-use DNA purification kits labs rely on today didn't exist. Neither did sequencing machines, which automated many complex lab processes. Instead, researchers had to manually do the reactions, make and run large gels, and read and record the data by hand. Saying anything meaningful about the sequences produced was almost impossible, because genetic databases and specialized search software tailored to genome work also didn't exist.

In time, advances led to the automated Prism sequencers that did the heavy lifting for Venter and Collins as they raced to complete the human genome in 2003. Decoding the first human genome took thirteen years, and the total cost of the Human Genome Project—which also included nontrivial expenses unrelated to genome sequencing—cost $3.2 billion. If Venter and Collins had started the project again in 2003, with all the new technology then available, they could have completed it for only $50 million in less than a year. By 2007, a startup built an even faster system that sequenced James Watson's DNA for $1 million. Just ten years later, a new generation of machines could read single molecules of DNA or RNA, which meant that researchers could peer into individual cells to see which genes are turned on or off. UK-based Oxford Nanopore Technologies makes a sequencing machine that's both the price of an iPhone and half its size. Astronaut Kate Rubins used it in 2016 when she performed the first successful DNA sequencing in space. Sequencers are getting smaller—and smarter.[27] Roswell Biotechnologies, headquartered in San Diego, is developing molecular electronic sequencing technologies that fuse DNA enzymes directly to semiconductor chips. Such chips will be able to record what individual enzymes are doing electronically; by doing so, they effectively wiretap their activities. The hope is that in the next year or two, a portable device will be able to sequence an entire genome in under an hour for less than $100.

The rate of progress in sequencing—from $3.2 billion and thirteen years at the beginning, to $100 and sixty minutes a short time from now, all in the space of thirty years—is unparalleled in any other industry. But that only encompasses reading the genome. For the really big stuff, we're just getting started.

CELLS ARE TRICKIER to program than a typical computer, in part because we don't have a complete understanding of the cell's machinery, and in part because biology is a water-based technology. This makes it different from technologies that are based on, say, silicon chips and electronics, where electrons whiz around on fixed paths while precise, high-speed switches control the flow. The cell is a vat of soup containing thousands of different molecules, and they are all constantly jiggling around and interacting, but moving very slowly compared to zippy electrons. Cellular processes and code aren't completely random, but they aren't linear and logical, either, which makes it difficult to predict exactly how any given biological system will behave. Cells and their components don't come with owners' manuals—they lack standards or specifications that would normally help an engineer build a device.

Traditionally, molecular biology experiments have been performed either in living organisms (*in vivo*) or inside of test tubes (*in vitro*). But since synthetic biology makes use of machine learning, experiments can be simulated on a computer. Welcome, then, to the era of *in silico* experiments. For example, a polypeptide chain with more than a hundred amino acids is considered a protein. There are more possible sequences in that tiny chain than there are atoms in the observable universe. But with *in silico* modeling, researchers can test how different genetic combinations might interact, learn to predict cell behavior, and run experiments to see what happens when biological processes continue to progress after a synthetic intervention.

Still, designs can only be tested so much *in silico*. Scientists also need to observe real-world biological activity, and for this they need to grow cells and build molecules. And they need faster and faster ways to synthesize DNA. Like our first computers, the first DNA synthesizers were highly trained people capable of completing challenging and repetitive tasks. But between the noxious lab fumes and the monotony of these tasks, chemists were eager to have this part of their work automated.

The first DNA synthesizer was introduced in 1980, when Vega Biotechnologies brought to market a machine the size of a microwave oven that could automate DNA production.[28] It cost $50,000—around $160,000 in today's dollars—and could make one DNA fragment (called an oligonucleotide, or oligo) per day, so long as it was only fifteen bases long. Since then, the cost of making oligos has come down dramatically, to pennies per base or less. Millions can now be synthesized at once.

Modern synthesizers are highly precise and can add the right DNA base to a chain with 99.5 percent accuracy. But increasing the cycles of adding bases to a chain increases the probability of introducing errors, given the vast volumes of bases involved. This limits the length of oligos that can be made with chemical synthesis to a few hundred bases, with most clocking in at around sixty bases. For many tasks, like the gold-standard PCR (polymerase chain reaction) test for COVID-19 that detects RNA specific to the virus, this size is sufficient. But to write a gene-sized fragment, or something larger, oligos need to be assembled into a longer chain. In nature, DNA synthesis and DNA assembly happen symbiotically. For now, these are separate processes in a lab. And a single mistake in genetic code would disrupt everything that happens down the chain.

Within the bioeconomy are companies founded to reduce some of this uncertainty. Twist Bioscience, to take one example, engineered a system that can churn out large volumes of DNA sequences with low error rates at very low costs. The company's technology uses a silicon chip that has tiny wells etched into it. The wells are filled with genetic material, and its DNA is then assembled into precise sequences. Compared to the conventional methods used in life sciences labs, Twist's innovation reduces the need for expensive reagents by a factor of 1 million, while increasing the number of genes that can be synthesized by a factor of 9,600.[29] Scientists can send DNA designs to Twist and receive DNA molecules a few days later. Before you get too excited, though, know that Twist doesn't sell to just anyone. You have to be a verified, authorized user, meaning someone attached to a registered academic lab or an approved business. They also won't synthesize just anything. Each DNA sequence is screened against a database of potentially dangerous sequences, like those of viruses or toxins. Not all DNA synthesis companies do this, however—nor, incredible as it may seem, are such screenings legally mandated. But more on that in a later chapter.

DNA SYNTHESIS TAKES place in settings like those of sci-fi movies set in the distant future: enormous white robotic arms following set paths on the floors of sterile, bright rooms. Those robotic arms traverse the space, picking up the plates and chips dotted with the tiny wells that are injected with different genetic materials. These robots work collaboratively, suctioning and dispensing liquids as the DNA molecules are synthesized and assembled from bytes of computer code and prepped for shipment.

Building and outfitting a modern biolab like Twist Bioscience can cost tens of millions of dollars. But smaller companies have other options. Biofoundries are facilities where high-throughput, liquid-handling robots and computer systems work in sterile environments to genetically engineer living systems, with computers logging all the activities and data. They're similarly expensive to build and maintain, but they perform work for hire. Biofoundries are synthetic biology's version of ghost kitchens, those shared commercial kitchens used by delivery-only restaurants operating out of large cities. Because resources are shared, biofoundries can perform lots of experiments at scale. Some foundries, including South San Francisco's Emerald Cloud Labs and Menlo Park–based Strateos, have embraced laboratory virtualization wholeheartedly, freeing scientists to program and operate them from almost anywhere.

Such biofoundries are being helped along by one unexpected but recognizable name: Microsoft. One of the company's research divisions works from an outpost in Cambridge (England, not Massachusetts), which was founded in 1997 and runs its molecular biology lab. It's a fitting location:

Watson and Crick made their DNA discoveries at Cambridge. In 2019, Microsoft launched a platform called Station B, with the idea of creating end-to-end interconnected applications and services for synthetic biology.[30] It's partnering with startups to develop the open-source programming language used for biological experiments, and other startups that automate the work of lab machines made by different manufacturers. For example, the platform is used to replace first-generation synthetic biology instructions, like "shake a test tube vigorously," with precise digital commands for lab robots.

Along with researchers at the University of Washington and Twist Bioscience, Microsoft is also exploring a new, unusual use for DNA involving information storage. DNA is already nature's hard drive. What if it could store other kinds of information, too? Today, you would find and store a picture of actor Dwayne "The Rock" Johnson on your computer by writing a complete file to memory. In the future, that same picture would instead be broken into thousands of tiny fragments and written into thousands of individual strands of DNA. Once sequenced, the DNA information could be reassembled by computers into the original file.[31] In 2019, the researchers prototyped the first fully automated read-write DNA storage system, using it to first write and then read the word "Hello" in just five bytes of data.[32] The catch? The turnaround time was twenty-one *hours*, not the milliseconds it would have taken a traditional computer to retrieve the file. It was an early example of a new kind of computer memory, where molecules were used to store data, and electronics were used for control and processing. With this working prototype in hand, Microsoft, Twist, the University of Washington, the sequencing company Illumina, and the digital storage company Western Digital founded the DNA Data Storage Alliance, aiming to create standards upon which a DNA storage ecosystem could be constructed. A dozen other member organizations quickly signed on.

It's clear why DNA data storage is an attractive proposition: the world generates a staggering amount of information each year. The optical, magnetic, and solid-state memories in use today cannot keep pace for much longer. Gargantuan amounts of information can be stored in a minuscule amount of DNA: just one gram can store over 200 million DVDs worth of information.[33] At this density, all of the world's digital information could potentially be stored in DNA molecules suspended in about nine liters of solution—picture two and a half gallons of milk—and, someday in the farther future, specific digital files retrieved as needed.

Advances in DNA synthesis have eliminated what used to be one of the highest barriers in genetic engineering: translating digital DNA code into code that a cell can run. But most synthesized DNA today is just a few thousand bases long, which is only about enough for a single protein. Building anything more complex, such as writing the complete genome of a microbe, requires tedious rounds of assembling fragments and precision sequencing before the biological design can be booted up, tested, and debugged. Biofoundries simplify or automate this drudge work but, like the first mainframe computers, they are expensive to set up and operate and have limited capacity. This isn't a problem for venture-capital-backed companies like Moderna, that pursue billion-dollar markets, but it does make it hard for most academic scientists to participate. That's why a smattering of research institutions around the world have built their own noncommercial foundries. In 2019, sixteen organizations banded together to form the Global Biofoundry Alliance to cooperate on these issues and address common challenges, such as locating the best prices for DNA synthesis, sourcing talent, and finding sustainable business models.[34]

Critical challenges still loom. For example, manipulating DNA during the assembly process is tricky, as there are lots of opportunities for things to go wrong. DNA fragments are delicate and can break, or there might be an invisible contaminant lurking in the lab. While some early standards exist, there aren't yet standardized practices between (or sometimes even within) labs for calibrating equipment, controlling processes, or using metadata.

BIOTECHNOLOGY IS ONE of the most complex industries in the world. And not just because of how cellular processes defy standardization, or the expense and precision that lab work requires. Intense regulatory frameworks, which require extensive testing, apply to any experiments performed on organisms that might be consumed, injected, or released into the natural world. The oversight ensures that the end product actually does what researchers designed it to do—and safely. The list of organizations involved would fill a dozen pages, and yet none of them use a single set of standards or agreed-upon rules.

Like all pharma companies, Moderna had to conduct a series of tests, beginning with a preclinical study, before it could deploy its vaccines to the general public. Initially, tests are performed in the lab on cell lines, which are regulator-approved cells that are grown and maintained for experimentation. The team needed to make sure that its mRNA vaccine successfully produced the correct non-harmful elements of the virus in the right places to trigger the body's immune response.

Moderna completed its vaccine candidate design (mRNA-1273) just two days after Dr. Zhang Yongzen and his team published the sequence of the virus. Then it made the gutsy call to move directly into clinical-grade manufacturing before any regulatory approvals had time to come through. They weren't trying to circumvent the regulators, but given the speed at which the virus could spread, and the potential for it to cause a pandemic, Moderna's decision makers wanted to get as much of a head start on the clinical trial process as possible.[35]

They didn't know it at the time, but a small German biotech company, BioNTech, with similar experience working with mRNA, was also engineering vaccine candidates, and it was forging partnerships with Pfizer in the United States and Fosun, a Chinese pharmaceutical company. By early February 2020, both Moderna and BioNTech had batches ready to move through the traditional, phased clinical testing process.

On March 27, 2020, Moderna's candidate began its Phase 1 study, which tests dosage and safety on a small group of human volunteers. At that point, the World Health Organization listed fifty-two additional vaccine candidates, the majority of which used inactivated or weakened whole viruses. Phase 2 trials, which test efficacy and side effects, began just a month later. Moderna's Phase 3 trial, which included thirty thousand participants, was designed to assess overall effectiveness and to continue testing safety. At the time of this writing, the mRNA vaccine was 94 percent effective in preventing symptomatic cases of COVID-19. (Phase 4 studies monitor the safety and effectiveness of a drug once it has been brought to market, scanning for very rare or long-term adverse effects that may not have been revealed in earlier trials.)[36]

When drugs are tested in the real world, they don't always work. There is always some chance that more tinkering will be necessary, or that it will fail completely. But this wasn't the case with mRNA-1273. In a December 2020 interview, conducted just hours before the FDA approved his company's vaccine, Bancel said that he was most proud that the vaccine was "100%, atom-by-atom," the one they'd made on a computer back in January.[37] It had worked perfectly out of the gate—a home run for synthetic biology. Even with Moderna's success in defending against this coronavirus, there are external factors to manage, such as monitoring production, to ensure that every batch is accurately formulated. Workers at a Baltimore-based facility encountered problems in mixing up some ingredients for Johnson & Johnson's vaccine, ruining 15 million doses.[38] And a packaging flaw in one batch of BioNTech vaccines delivered to Hong Kong and Macau forced the suspension of local vaccine drives in March 2021.[39]

Moderna and BioNTech are increasingly confident that, if required, updated vaccines to manage new variants can be created even faster than they were in 2020. But, importantly, regulators now have real-world experience with synthetic mRNA vaccines, which should clear a new pathway for other use cases and clinical trials. Moderna has nine mRNA vaccines in their development pipeline, and some are already in Phase 1 trials. Meanwhile, researchers at Yale University, in partnership with Novartis, are preparing to test an mRNA vaccine for malaria that has a big technological upgrade: the RNA can self-replicate in the body, meaning that a much smaller dose will be required, thus making it easier for the company to produce millions of doses.[40] As a bonus, it won't need the super-cold freezers required by mRNA vaccines for COVID-19.

A 1965 ARTICLE with a simple promise—"Cramming More Components onto Integrated Circuits"—changed the course of modern computing.[41] Written by Intel cofounder Gordon Moore, the article laid out his theory that the number of possible transistors that could be placed on an integrated circuit board for the same price would double every eighteen to twenty-four months. This novel idea became known as Moore's Law, and it very quickly became the fuel for a nascent but flourishing computer industry, because it validated the bold visions of those early innovators. The confidence that Moore's Law–powered forecasts gave investors emboldened them to divert resources into computing, and it empowered company leaders to plan ambitious new products and services as the value network for computing developed. While no company could ever schedule its R&D breakthroughs, businesses now had an indicator about when the timing and conditions would be right to make big improvements and daring new initiatives. References to Moore's Law became commonplace in computing, but at the time Moore's paper was published, few outside of Silicon Valley had heard of him, or the company he'd cofounded. Those were very different times: when Moore published his article, the term "Silicon Valley" didn't even exist yet.

Today, there is a Moore's Law analog in synthetic biology, and it's named for a physicist at the University of Washington named Rob Carlson. In the early 2000s, Carlson was studying the rates at which different biotechnologies were improving. Inspired by Moore's paper, he made the case, in a 2010 book, *Biology Is Technology*, that as technology improved, the cost of sequencing and synthesis would sharply decline. So far, his calculations—which have come to be known as the Carlson Curves—are holding true. According to the National Human Genome Research Institute, the cost to sequence a high-quality draft of a human genome in 2006 was $14 million, and a finished sequence cost between $20 million and $25 million. By mid-2015, the cost for a finished sequence had dropped to $4,000.[42] Today, BGI, a Chinese company, can sequence a genome for $100. That's a human genome for less than the price of a pair of Air Jordans, and it is why millions of people's genomes are being fully sequenced each year.[43]

As genome sequencing gets cheaper yet, it promises to fuel an explosion in new, inexpensive diagnostic tests, such as early-detection cancer screenings. It could also make it feasible for every baby to be sequenced right in the delivery room, or even in utero or at conception, and it could open up a new world of personalized prenatal testing for traits like future mental performance. In just a few decades, it could unlock the code of every plant, animal, microbe, and virus with which we share our world. All this data is raw material for a new generation of genetic engineers who, empowered with advanced software tools, ever cheaper and more powerful synthesis technologies, and cloud laboratories, will only need laptops and funding to realize their creations. This portends a coming genetic divide if testing isn't equitable, and a potential for genetic enhancement by those who can afford the most advanced biotechnologies.

What will the bioeconomy bring? As with much in synthetic biology, Craig Venter may have glimpsed this future before anyone else. Once the Human Genome Project was completed, and after his falling out with Celera, he turned his attention to writing genomes, founding Synthetic Genomics, Inc., in 2005, near the beaches where he loved to surf in La Jolla, California.

In 2017, Synthetic Genomics demonstrated a sort of biological printer, which Venter called a digital-to-biological converter, or a DBC. It consisted of a robotic DNA/RNA synthesizer-assembler system about the size of a sofa. Researchers sent various genetic programs to the DBC and printed out DNA for a protein, an RNA vaccine, and a bacteriophage (a virus designed to infect bacterial cells). Each product was then manufactured in their lab. In other words, their DBC had essentially allowed them to shrink their company—indeed, all of the design and manufacturing phases of synthetic biology—into a box that, in another setting, would fit nicely into any living room.[44]

Today, the company, which is now called Codex DNA, is working on a powerful foundry even smaller than the original—a box the size of a cooler you'd take to the beach. Instead of incorporating a synthesizer, the new machine uses proprietary cartridges, much the way an inkjet printer operates. To make a sequence, scientists would upload their requirements and order cartridges preloaded with pieces of their desired genetic code. A few days later—shipping the cartridges would likely take longer than manufacturing them—the cartridges would be loaded into the machine, and a few buttons would be pressed for the experiments to begin. With a generator and satellite internet connection, scientists could theoretically build DNA in the middle of the Brazilian rainforest. Or on a battlefield. Currently, DARPA is supporting the development of even smaller devices that can manufacture drugs and therapies in the field, in order to protect troops against novel biological threats. Imagine a situation in which soldiers are exposed to an engineered pathogen, or a new and dangerous naturally occurring virus. At that point, genetic sequencing could quickly determine the genome of the pathogen. A new vaccine or medicine could be quickly designed and then downloaded and printed to treat wounded or ill soldiers. Perhaps that same machine, or another connected to it, would then manufacture doses needed to protect force readiness and keep troops healthy.

If, in the future, organisms and medicines can be faxed around the planet, why not send them to other planets? Venter worked with NASA to demonstrate that if bacterial cells based on DNA were discovered in Martian soil, they could be sequenced on site, and the data could be beamed to labs on Earth for reconstruction and testing. The system could also work in reverse. With a biofoundry on Mars or the moon, sending crucial supplies, plants, and animals would be as easy as sending an email. In other words: a hyper-upgraded "speaking telephone" that uses "extra-terrestrial relays" as part of a "galactic network" for the purpose of biological teleportation.

Until then, the value network for synthetic biology's future continues to progress today. As you are about to see, a vast array of materials, pharmaceuticals, textiles, plants, and animals, for this planet and for others we might someday explore, are already in production.

| SIX |

# THE BIOLOGICAL AGE

IMAGINE A WORLD IN WHICH SHRIMP ARE PRODUCED IN A LAB. ONE where they're not brought to market by ships trawling (and destroying) the sea floor with giant nets, thus leaving all the other fish and aquatic life typically ensnared in those nets blissfully undisturbed. Imagine if fruit were grown indoors, all year round, just beneath the grocery stores that sell them. Instead of harvesting berries before they are fully ripe, blasting them with temperature-controlled washes to kill pests, sealing them in shipping containers, and sending them halfway around the world, you could pick what you wanted right off the bush.

Or imagine a future in which healthy living was about upgrades and optimization, rather than restriction. One where hangovers would become a relic of a bygone era, because, before enjoying a night out, you'd eat a special probiotic supplement to prevent alcohol's notorious aftereffects. Imagine a world in which dieting was unnecessary, because a biometric test would reveal your body's metabolic levels, food sensitivities, and other data, so you'd know which foods to eat, what to drink, and when. Imagine a world in which genetic diseases that cripple and kill infants, such as sickle cell anemia and muscular dystrophy, would all be prevented before birth.

Now think back to all the screens you've cracked, nails you've chipped, and lenses you've destroyed in one way or another. Imagine the day when biological coatings ensured that surfaces would quickly heal themselves, regardless of the damage inflicted upon them. No matter how many times you misjudged exactly how to maneuver in and out of your garage, the finish on your car would still look pristine. Organic nail polish would dry without the need for harmful UV lights or harsh chemical topcoats.

The foundation for these alternate realities already exists. In fact, some of these innovations are already moving from the fringe to the mainstream as the bioeconomy's capabilities come into focus.

Many key phases of human progress have been described in terms of the materials we used to transform our built environments. During the Stone, Bronze, and Iron Ages, humans created the basic technologies for everyday tools, agriculture, construction, and warfare. Later, we harnessed glass for ornamentation, bottles, windows, lenses, and medical equipment. Steel enabled us to build skyscrapers. Plastic made the mass production of disposable containers possible, which led to global supply chains for food, medicines, and water, among other products. Our current moment—in which we are learning to manipulate molecules, engineer microorganisms, and build biocomputing systems—is the start of a new era in civilization's evolution: the Biological Age. What we build during this new age will unlock new business opportunities, mitigate or even reverse environmental damage, and improve the human condition in countless other ways—both on Earth and in off-planet colonies.

Synthetic biology will transform three key areas of life: medicine, the global supply of food, and the environment.

## MEDICINE

In the next two decades, synthetic biology technologies will be harnessed to eradicate life-threatening disease and to develop personalized medicines for individual people and their specific genetic circumstances. Researchers will genetically engineer viruses to treat cancer, and they will grow human tissue in a lab for organ transplantation and to test new therapeutic treatments. New technologies will monitor us continuously, eliminating traditional doctor's exams. Most importantly, we will engineer healthier people: predicting and eliminating genetic disorders, and potentially making enhancements, to babies before they are born.

### *Eradicating Disease*

Before a short vacation to India, a doctor made sure her wife and two children were all up to date on their shots: polio; measles, mumps, and rubella; diphtheria; chicken pox; flu; hepatitis A and B; tetanus.[1] They weren't traveling to an area with cholera or yellow fever outbreaks. They were staying in New Delhi and planned to make day trips to various tourist sites. Watching the sun set over the Taj Mahal in Agra was magic: as the light gradually faded, it cast a vivid orange-pink filter over the bright marble buildings. A perfect day, but for the swarm of mosquitoes that thickened as the colors changed.

A few weeks after the family came back home, one of the moms—the doctor—felt as if she'd gotten the flu. She was stricken with chills, fever, and body aches, so she drank some water, took some Advil, and went to bed. But her condition quickly deteriorated. By the time she arrived at the hospital, her blood pressure was falling. The triage nurse asked a long list of questions: Had she taken any new medications? Did she have any allergies? And finally, had she traveled out of the country recently? As soon as the doctor mentioned that night at the Taj Mahal, the nurse called for an infectious disease specialist. It turned out to be a very good call: the doctor had contracted malaria, the deadly disease carried by mosquitoes.

Mosquitoes use the sharp tip of their needle-like mouths to pierce your skin and leech blood from your body as they also inject an anticoagulant to prevent your blood from clotting. Only female mosquitoes have this specialized set of biomechanics; male insects feed on plant nectar. Since females feed on animals and people, female mosquitoes are key vectors for disease. They spread malaria, dengue fever, and a bunch of viruses: West Nile, Zika, chikungunya, and Eastern equine encephalitis, among many others. What's the most prolific and deadly predator on Earth? It isn't a snake, shark, or scorpion. It's not a bear. It's not even humans. It's the mosquito.

Each year, malaria kills more than four hundred thousand people, most of whom are young children.[2] The disease isn't caused by a virus or bacteria, but rather an organism called a plasmodium. Plasmodia are clever, shape-shifting organisms adept at evading immune systems, which is why they spread and persist. The only vaccine that exists for malaria requires four injections and is relatively ineffective: it only provides temporary resistance. Someone infected with malaria can become reinfected over and over again. The best defense we currently have against the disease is early diagnosis and treatment. That a doctor misdiagnosed her own case of malaria tells you something about how dangerous malaria is—and why mosquitoes became an early target for synthetic biology.

Because we cannot easily get rid of disease-carrying mosquitoes—they breed quickly and are hard to capture—we've instead spent decades developing lotions and sprays to ward them off. The US Army used DEET as a repellent after World War II, but DEET is toxic. (When mixed incorrectly, it can melt plastic.) Besides, some mosquitoes have developed a genetic resistance to it.

But there is now a way to prevent malaria from spreading without having to kill trillions of mosquitoes. In 2021, geneticists at the Imperial College of London used a "gene drive"—a genetic modification that results in a majority of offspring carrying a desired trait—to deal with this disease. They relied on the gene-editing technology of CRISPR—that editing technique to cut a specific site in DNA—to modify the sexual development and other traits of female mosquitoes. Females born with the edited genes have different mouths, so they can't bite, and they also can't lay eggs, which means they can't spread the malaria parasite. Without the gene drive, the mutation would have spread too slowly through the population; with it, nearly 100 percent of offspring inherit the new mouth design. Gene drive technology is powerful—and permanent.[3]

New mosquitoes with different genetic modifications are being developed and tested at scale in a high-security facility in Terni, Italy, and elsewhere.[4] In 2021, millions of other genetically engineered mosquitoes were scheduled for release in the Florida Keys to curb the spread of Zika. The Florida Keys Mosquito Control District Board of Commissioners had approved a pilot project to introduce genetically edited male mosquitoes that pass on a gene that makes it difficult for their offspring to reproduce.[5]

Local authorities, who have been dealing with steadily growing cases of dengue fever and West Nile virus, believe that a smaller mosquito population will curb the diseases—and stop them from needing to douse the Keys with insecticides or poisonous chemicals.

### *Personalizing Medicine*

CRISPR allows scientists to edit precise positions on DNA using a bacterial enzyme. Roughly eight thousand diseases originate from just a single gene. Such monogenic, or single-gene, disorders were difficult to diagnose and treat before we had DNA sequencing. They include sickle cell anemia (a heritable condition that causes red blood cells to contort into a sickle shape) and cystic fibrosis (which causes the body to produce a thick, sticky mucus that clogs the lungs and digestive system). But CRISPR could correct these gene mutations, and the cell's DNA repair machinery could restore edited cells back to health.

Even more people suffer from polygenic diseases, which are more complex to manage because they stem from the combined actions of more than one gene. Coronary heart disease and atherosclerosis are polygenic conditions. So is hypertension, which in some people is heritable. Amy's father is one of those people. He was diagnosed with severe hypertension in his early twenties, when his blood pressure suddenly went haywire: seemingly out of nowhere, his nose started bleeding, and he became so dizzy that he fell over. He was otherwise healthy and active, had never smoked, and had avoided alcohol because of his religious upbringing. But the first time he displayed any symptoms left him in a life-threatening situation. For the past fifty years, he's been under the care of specialists at the Cleveland Clinic and Johns Hopkins Hypertension Center, who regularly tinker with the dosages of the complex cocktail of prescription medications he takes: he must swallow twenty-seven pills throughout the day, every day, to stay alive. The job of his physicians now is to figure out how to counteract the side effects of all those meds on his body, with additional medications that also weren't designed specifically for him. A conservative estimate of his medical expenses, between all those prescriptions, doctor visits, and occasional trips to the emergency room, is $50,000 annually, which adds up to roughly $3 million in health-care costs during his lifetime—and that's before you start adjusting those costs for inflation. He's in a privileged situation: he has excellent insurance, access to world-class research hospitals and physicians, and a supportive and helpful family. Still, synthetic biology techniques could, in the future, target and edit or rewrite the specific portion of the genome responsible for his hypertension.

What about all the other diseases that require medication, but won't respond to CRISPR editing? In the future, therapeutics could be made as needed, rather than manufactured at a global scale. As we saw in the previous chapter, portable sequencing machines can already determine the presence of a virus or bacteria, and the technology to produce real-time medications isn't far off. If you were able to sequence a genome quickly and had the right materials to synthesize a treatment, you could print out a protective barrier to defend against any number of pathogens. Imagine a kit that included freeze-dried molecules that were engineered for later insertion into a cell. The molecule would remain dormant until you needed to use it, like the dried beans and mushrooms you might use to make soup. Activating the treatment would require adding just the right amount of water to rehydrate and bring the biological system back online. Once up and running, you would then add engineered DNA instructions for vaccines or antibiotics. Hikers, athletes, soldiers in the field, and school nurses could carry a kit the size of an iPhone and mix medications whenever and wherever they were needed.

### *Beating Cancer*

Amy's mother developed a rare form of neuroendocrine cancer in her mid-fifties. Her cancer didn't have an obvious culprit. She didn't drink or smoke, and aside from being overweight—she had a sweet tooth and loved French fries—she was vibrant and healthy. The symptoms, too, were unusual: her skin seemed to turn yellow overnight, and she lost weight quickly. Her primary care physician referred her to a team of specialists at the University of Chicago, who then referred her to the MD Anderson Cancer Center in Texas. She had a cancerous tumor, but that wasn't the problem. Doctors had detected worrisome cells far away from that site, with traits similar to both nerve and hormone cells. They showed up at her pancreas and in her lungs, and they were aggressive and replicating quickly. When the DNA of a neuroendocrine cell mutates, it seems to cause tumors to grow anywhere spontaneously. Which meant that this one visible tumor likely wasn't the only one already growing inside her body.

She fought valiantly, insisting that life would go on as planned. She covered the chemotherapy port in her arm with long sleeves, found a wig that matched her trademark pixie haircut, and continued teaching her fourth-grade class. Because physicians could not find the primary site where cells were going bad, they were forced to combine chemotherapy treatments designed for other cancers. She spent Saturdays connected to IV bags, under warm blankets, without the wig, without the brave face she wore everywhere else. The first four hours were spent connected to an IV bag designed to treat pancreatic cancer, and the second four with a bag designed to treat lung cancer. She would finish six or eight weeks of treatments, take a few weeks off to regain her strength, and begin another round. Chemo was both keeping her alive and drastically shortening her life. After a year she decided, with family support, to enter hospice care. She was gone a few days later.

It would be difficult to develop a single, universal vaccine for cancer, because each cancer is unique, and cancer isn't a single disease but rather a catch-all term for a constellation of more than a hundred known genetic mutations. Perhaps that's why we tend to name cancers by location rather than mutation: lung cancer, bone cancer, neuroendocrine cancer. But individualized vaccines could be developed to defeat the mysterious cancer that killed Amy's mother, with genetically engineered viruses designed to target and kill cancer cells. Another option: using mRNA to spur the body to build up its immunological defenses to find and kill cancers, while also protecting against them in the first place. One way this can be done is through CAR T-cell therapy, which is short for chimeric antigen receptor cell transfer. In this treatment, the T-cells (specialized white blood cells) are spun out of a patient's blood, modified in the lab, and reinjected to fight off the cancer cells.

Long before they were making COVID-19 vaccines, both Moderna and BioNTech were researching immunotherapies for cancer. After analyzing a tissue sample from a cancerous tumor, the companies ran genetic analyses to develop custom mRNA vaccines, which encode protein-containing mutations unique to the patient's tumor. The immune system uses those instructions to search and destroy similar cells all throughout the body, which is similar to how the companies' COVID-19 vaccines work. BioNTech is currently in clinical trials for personalized vaccines for many cancers, including ovarian cancer, breast cancer, and melanoma. Moderna is developing similar cancer vaccines. Both companies understand that the world's most powerful drug manufacturing factory on Earth may already be inside you. We just need to figure out all the ways we can harness it.

### *Customized, Lab-Grown Human Tissue*

mRNA cancer vaccines or custom viruses would need to be tested rigorously, and in an ideal world, quickly. Such clinical trials are costly and require a lengthy regulatory review. In the absence of an emergency like COVID-19, it can take a decade or longer to get all the approvals. It's also difficult and dangerous for scientists to study how living human tissue responds to viruses and medications. Brain or heart tissue, for example, can't be excised from a living person. We need a way to make it easier to test new treatments and reduce the time it takes to develop them.

Synthetic biology has some solutions for these problems. It offers us, for example, the ability to engineer and grow organoids—tiny blobs of tissue grown from human stem cells. As we write this, lab-grown lung and brain tissues are being used to research the lasting effects of SARS-CoV-2.[6] Miniature guts and livers are also being grown, and infected with the virus, in high-security labs. The Wake Forest Institute for Regenerative Medicine is leading a unique $24 million federally funded project to develop a "body on a chip," which will include different combinations of organoids.[7] Picture a computer chip, but with a transparent circuit board that's connected to a system pumping a blood substitute through it. With it, researchers can poison a mock respiratory system with new viruses, lethal chemicals, or other toxins to see how the body would react, and then test potential treatments on living human tissue without harming humans or other animals.

Miniature snippets of the nervous system are being used to create miniature blobs of brain tissue. In 2008, researchers created the first cerebral organoids, which provided a better understanding of certain brain functions. Cerebral organoids have since been used to research autism and other diseases, such as the Zika virus.[8] When joined together, lab-grown muscles and cerebral organoids can establish neural highways and process information. Researchers at Stanford University are experimenting with self-assembling tissue, which they dubbed "assembloids," that respond to stimuli.[9] Starting with human stem cells that sew themselves together in a petri dish, the researchers developed a working prototype of a nerve-cell circuit that represents the cerebral cortex, spinal cord, and skeletal muscle. When they stimulated the cortex cells, the message was transmitted all the way to the muscle, which twitched in the lab dish. In another study, researchers created human forebrain organoids—the forebrain is the part of the brain responsible for thinking, perceiving, and evaluating our surroundings. Research is underway elsewhere that would transplant bits of human brain organoids into rats, which raises both complex ethical concerns and, perhaps, fears of super-rats that process information as well as humans. But more on that in the next chapter.

### *Doctorless Exams*

When deCODEme's genetic testing kit first hit the market in 2007, it promised to assess consumers' genomes for disease risk and origins. The Iceland-based company then charged $985 for a screening.[10] That same year, Google-funded 23andMe began offering genetic testing for $1,000 and returned results in a few weeks, along with access to an online Genome Explorer dashboard.[11] Both raised concerns among genetic counselors, the specially trained health professionals that advise would-be parents and people diagnosed with genetic disorders, who thought that customers would gain access to their genetic risks without having sufficient context to interpret the test results. Initially, 23andMe was only authorized to ship its kits to a handful of states, as laws in Maryland and New York, among others, prohibited residents from buying any direct-to-consumer tests that provided health information. In 2017, the FDA finally allowed 23andMe to market its tests for ten conditions, which included markers for Parkinson's and Alzheimer's.[12]

But consumer tests aren't just about genetic fortune-telling. The human microbiome is a mini-universe of the genetic materials living on and inside of our bodies, which we inherit from our gestational mothers. The number of genes in the bacteria, fungi, protozoa, and viruses that make up our microbiomes is two hundred times the number of the genes in the human genome. That microbiome weighs nearly five pounds, and it lives mostly in your gut and on your skin. Microbiomes differ greatly from person to person, even if they're siblings who live in the same city. How well you digest lactose, how vulnerable you are to skin cancer, how well you sleep, your probability of developing anxiety or becoming obese—all of these traits are linked to the microbiome and influenced by what you eat and drink, whether you smoke, what chemicals your body comes into contact with, and what medications you take. That data once was collected over several visits to an allergist, but today, at-home tests can determine the genetic makeup of your microbiome. Some companies will mix together special probiotic compounds to mitigate conditions or optimize the symbiotic relationship your body has with all those microorganisms.

Which means that the next frontier in medicine could be doctorless exams. You won't have to worry about making appointments to visit a diagnostic lab, and then wait in lines at the lab and wait even longer for results, because a new crop of technologies will analyze your data at home. Such technologies will start with the everyday equipment in your bathroom, which you'll soon use to collect biological samples for daily monitoring and testing. To be blunt about it, our toilets already come into direct contact every day with two key sources of data—our skin and our waste—so they offer an excellent route to monitoring our real-time health. That's what gave Stanford University researchers the idea for a diagnostic toilet, albeit one outfitted with the sort of equipment you don't typically want inside a bathroom: cameras, microphones, pressure sensors, tiny robotic arms, motion detectors, infrared sensors, and a computer system equipped with computer vision and machine learning.[13] Their hunch was that regular diagnostic samples could be used as an early warning system for gut disorders, liver and kidney diseases, and cancers. It turned out to be correct. If you were following the Stanford experiments, Toto's "Wellness Toilet" shouldn't have come as a surprise when it was announced at the Consumer Electronics Show (CES) in 2021. As far-fetched as this may sound, this is a real device intended for everyday use: the high-tech toilet uses a similar array of sensors to analyze "key outputs" and provide users with insights on hydration and their diets. There are portable kits, too: the at-home test startup Healthy.io has developed a urinary tract infection test kit that uses a mobile app to connect patients who have positive results with an online doctor and sends a prescription to a nearby pharmacy if needed. Healthy.io also partnered with the National Kidney Foundation to offer an annual kidney test kit to detect early signs of disease.

In the near future, diagnostics will require even less effort, as wearable and ingestible devices will provide data for remote health monitoring (RHM). Phones and wearables have long been collecting and interpreting such data; consider how people who wear an Apple Watch know that an unusually high or low heart rate or irregular rhythm may suggest atrial fibrillation. Smartphones and smartwatches now take blood pressure readings and perform electrocardiograms, using apps approved by the Food and Drug Administration. RHM uses a network of digital technologies, the internet, and the cloud to collect medical data from patients and then transmit it for remote assessment by their health-care providers. Lots of data—heart rate, electrocardiograms, blood pressure, blood oxygen levels, kidney function, and more—can be mined to manage cases offsite. RHM can keep older people in their own homes longer, and it can reduce the number of in-person visits to clinics and hospitals. Ingestibles, which are tiny, pill-sized computers outfitted with sensors, cameras, and transmitters, can collect data from inside your body and beam it out to an AI-powered system for analysis. Researchers at MIT have developed an ingestible-based bacterial-electronic system to monitor gut health.[14] Other ingestibles can detect bleeding or tissue abnormalities, or even check to make sure a patient is taking prescribed medication.

### *The End (and Rebirth) of Medicine*

Manufacturers update the prices of medications semiannually, and drug prices are continually rising. During the January 2019 update, for example, 468 drugs increased in price in the United States by an average of 5.2 percent; in the January 2021 update, 832 drugs increased an average of 4.5 percent.[15] Drug manufacturers argue that patients don't see those increases at the pharmacy when they're picking up prescriptions, because insurance plans pay for the added costs. But list price changes have significant downstream effects on the overall cost of health care in America, with prices that continue to skyrocket. Insurance costs have increased a staggering 740 percent in the past thirty years. More than half of Americans rely on their employers for health insurance, which subsidizes those costs. Employer-based insurance for an average family costs businesses $20,576 a year, but such insurance still leaves families on the hook for a lot of expenses, including deductibles and medications that aren't covered.[16]

Expensive medicine is an addressable problem. One easy fix would be to empower people to know their baselines for key physical metrics, thanks to diagnostic equipment in their own homes. When a patient visits a doctor in Japan, she is asked for her normal temperature, rather than the doctor assuming it's 37°C (or 98.6°F). Many people are a few tenths of a degree away from that average number. According to the Centers for Disease Control and Prevention (CDC), only a temperature of 100.4°F and above is considered a fever, but patients with normal (for them) temperatures of 97.9 and 99.0 would have far different physical responses to a fever of 100.4. Aside from body temperature, there are thousands of other data points swirling around the body that could indicate anomalies relative to any individual person. Sensors, wearables, and ingestibles would feed data to machine learning systems, determine whether someone is deviating from their baseline, and provide people with actionable information.

An intervention could be as simple as drinking a glass of water, or, for more serious conditions, perhaps we would harness our own internal drug factories, using custom microbes and biological code—rather than swallowing twenty-seven pills a day, or suffering through an ad hoc combination of chemotherapies never intended for your particular cancer. This would eliminate the need for expensive drugs that were designed for the masses rather than your unique biological circumstances. It would also challenge the mechanics of our current pharmaceutical and health insurance industries. With enough public trust and acceptance, it could end medicine as we know it, which is expensive, unevenly distributed, and inaccessible to many. We could transition to a system of personalized medicine, with better health equity and outcomes for everyone.

But the ultimate prevention strategy for genetic disorders, diseases, and cancers linked to heritable genes is to predict, detect, and avert them before we are born. Some people opt in to genetic screening tests that can be performed even before pregnancy to determine whether the parents are carriers of disease. Tay-Sachs, a rare and fatal genetic disorder that destroys nerve cells in the brain and spinal cord, occurs more frequently in people with Ashkenazi roots. Prospective parents whose ancestors were Jewish and emigrated from Eastern or Central Europe can take an "Ashkenazi Panel" to determine whether they are carriers for the disease. Other markers can also now be tested. In vitro fertilization typically results in multiple embryos, and some people employing that technology choose to screen embryos for conditions such as Down syndrome or cystic fibrosis before implantation.

Researchers are developing a new technique that might someday enable people to upgrade their children before birth. Using algorithms to understand the tiny variations in DNA—single nucleotide polymorphisms, or SNPs—they hope to make accurate gene-based predictions about an individual's future.[17] If SNPs were read in vitro, before embryos were implanted, they could reveal whether that genetic combination had a higher probability of developing heart disease or diabetes. If an embryo was edited using CRISPR, embryos could also be optimized with the best possible traits, given the raw genetic material. Theoretically, parents could influence myriad traits for their offspring, including hair texture, resistance to a virus such as HIV or protection against Alzheimer's disease. This intervention, like the gene drive edit in mosquitoes, would have a permanent, heritable effect. It could eradicate certain diseases passed from parents to children, and in the process improve the entire gene pool.

Depending on where you stand, this will sound somewhere between "really exciting" and "gravely concerning." We will address existential risks in the next chapter. So far, at least, caution and ethical concerns have the upper hand.[18] A dozen countries have banned germline engineering in humans, though their ranks do not include the United States or China. The European Union's Convention on Human Rights and Biomedicine says tampering with the gene pool would be a crime against human dignity and human rights.[19] But all these declarations were made before it was actually possible to precisely engineer the germline. Now, with CRISPR, it is possible.

Many other possibilities that once would have strained credulity now loom. An emerging technology called in vitro gametogenesis, or IVG, will soon allow same-sex couples to create a baby using their own genetic material without requiring donor eggs or sperm.[20] A Japanese scientist, Shinya Yamanaka, won the Nobel Prize in 2012 for his remarkable discovery: a method to turn any cell in the human body into induced pluripotent stem cells (iPSCs for short), which can be reprogrammed with the functions of any other cell. Researchers at Kyoto University used that technology in 2016 to transform iPSCs from a mouse's tail into eggs, and eventually, baby mice. With this technology, our definition of genetic "parents" will shift drastically in the next few decades, from our current one-father-one-mother construct to parenthood in many different permutations. LGBTQ couples will be able to reproduce more easily using their own genetic material, without the need for donors. A woman who chooses to have children would no longer need a man's donor sperm: eventually, the technology will allow her to conceive a baby using her own genetic materials, and no one else's.[21]

And if a man or female-identifying person wants to become a parent someday? IVG will produce an embryo, but bringing that baby to full maturity would still require a gestational carrier. Historically, that has always been a woman; however, researchers at the Children's Hospital of Philadelphia created an artificial womb, called a biobag, which they used to keep premature lambs alive and developing normally for twenty-eight days.[22] In March 2021, a team of Israeli scientists grew mice from embryos in a completely artificial womb and kept them alive for eleven days.[23] We are still years away from synthesizing and growing a full-size organic womb, but the biobag represents a gestational carrier intervention that could help the thousands of premature babies born every year—and portends an eventual future in which humans may not need to carry their pregnancies at all. Within a generation, our conception of a "nuclear family" could look radically different, and far more inclusive, than it does today.

## THE GLOBAL FOOD SUPPLY

The Hoover Dam once held back the mighty waters of the Colorado River, but today chronic overuse by cities and farmlands, combined with epic droughts and soaring temperatures, has created a water shortage. The Colorado River's largest reservoir, Lake Mead, is only 37 percent full as of this writing, leaving millions of acres of farmland vulnerable.[24] The modern systems we use to cultivate food are contributing to the Earth's climate and ecosystem destabilization. Synthetic biology offers alternatives to resource-intensive farming and ranching and to the cold chains we rely upon to move perishable foods around the world.

### *Stabilizing Agriculture and Aquaculture Systems*

A few years ago, Amy and her Future Today Institute team were leading a scenario planning exercise for a company that made a popular frozen food. The majority of that product's key ingredient came from just one farm in Eastern Europe, where extreme weather events were becoming more common. That, coupled with bouts of civil unrest and growing nationalism, resulted in strikes, particularly at this farm, because workers knew the products were primarily destined for more affluent markets outside their own struggling country. There were other complexities: the product was perishable, and it had to be transported from the farm to a distant factory in Western Europe for cleaning, preparation, and processing into a frozen food for global distribution. The company typically met its targets, but a recent spate of extreme weather, which resulted in a catastrophic drought, decimated the crops required to manufacture its product. It became unable to supply it in the volumes the market demanded, and a successful marketing campaign drove consumers to empty shelves. But given the tenuousness of its supply chain, it was a miracle that a breakdown hadn't happened sooner.

Indeed, it's a miracle that our current agriculture and aquaculture systems work at all. Emerging weather problems, including fires, droughts, and extreme heat and cold, are becoming increasingly common everywhere, yet still very hard to predict. The politics of agriculture, too, are confounding. Between 2012 and 2021, the US government repeatedly changed course on the H-2A federal visa program for guest farm workers. Crackdowns on migrant undocumented farm workers subsided, then intensified, and now appear to be easing again. The agricultural workforce has been unstable as a result. And then there are the weird glitches that sometimes show up as we continue to test the physical limits of our supply chain systems—such as relying on enormous cargo ships that technically fit through the Suez Canal, but only under exactly the right circumstances.

Roughly one-third of the food produced every year for human consumption—1.3 billion tons—is wasted or lost.[25] In the United States, there is more food waste in landfills than any other material.[26] There are many reasons why. Chain restaurants wanting to ensure optimal freshness will discard food after a set number of hours, even if it is still safe to eat. Sometimes shipping delays lead to rotten produce. Fresh produce, dairy, and meats aren't just nourishment, they're marketing: we expect apples to be uniformly colored, carrots to be perfectly long and straight, and our factory farm eggs to be bright white with yellowish yolks, so misshapen products tend to rot before they sell. Enormous portions and extensive menu choices result in uneaten meals and leftovers at restaurants. When some chain restaurants close at night, they throw away any leftover products they've already made, because employees aren't allowed to take it home. Grocery stores and chains are reluctant to donate leftover food to organizations that serve those in need because of legal liabilities. In total, more than 40 percent of food waste happens at the retail and consumer levels in the industrialized world.[27] In developing countries, losses happen during harvesting, storage, and processing: machines break, unskilled laborers make mistakes, and razor-thin margins prevent contingency planning in case of real problems. This is where you get disease outbreaks, such as listeria, which leads to mass recalls.

The agricultural sector has been genetically tinkering with food for decades. The first genetically engineered bacterium was developed in 1973, followed by an engineered mouse in 1974, and a tobacco plant in 1983 that contained genes from another living organism.[28] These were precursors to the FDA in 1993 allowing companies to sell genetically modified seeds; a year later, the Flavr Savr tomato, which stays ripe longer than other varieties, was approved for sale in the United States.[29] That first wave of genetic engineering resulted in what came to be known as genetically modified organisms, or GMOs, which included modifications that are proprietary and intended for use in conjunction with particular herbicides and pesticides. Today, around 14 percent of all cotton grown worldwide is genetically modified, and nearly half of all soybeans are modified, too.[30] In the US, these figures are much higher, with GMO cotton and soybeans both above 90 percent.[31]

But the future of genetic modification will look very different from what we are seeing today. Consider the artificial leaf, which was developed at Harvard. It is a lab-created device that harnesses solar energy. When connected to a strain of bacteria, it converts atmospheric CO2 and nitrogen into organic forms that are beneficial to living organisms. Those hungry, solar-fed bacteria essentially overeat, to the point where 30 percent of their entire body weight is excess energy—stored CO2 and nitrogen. These microbes then get mixed into soil, and near the roots of plants they release all that nitrogen, which acts as an organic fertilizer. Once there, they also release the CO2, which remains trapped underground. The result: enormous crop yields without the environmentally poisonous side effects typically associated with chemical fertilizers.[32]

Turning bacteria into plant food, using CRISPR to improve seeds, improving plant-based proteins, and synthesizing lab-created meats will transform agriculture as we know it, and it will lead to more of what we eat being grown indoors. Expansive plant factories will grow crops genetically designed to use fewer resources, cause less damage to local environments, and produce greater yields. They'll also engineer produce and meat to enhance flavor and the concentration of nutrients. It will give us security in the face of an uncertain future, where climate change increasingly threatens the global supply of food.

It will all be a big change to how we grow and produce food, but we may not have a choice. Today, food insecurity affects one out of every four people on Earth.[33] The global population is projected to increase by two billion people by the year 2050.[34] We will either need to curb population—a very complex and fraught task—or we must increase our global food supply and distribution to all people. Simply growing more rice or raising more cows won't help: animals raised for meat, eggs, and milk account for 14.5 percent of all greenhouse gases.[35] Increasing our current food supply to meet future demand will further challenge our planet's climate.

### *Editing Livestock (and Agriculture)*

Outbreaks of African swine fever decimated the global pig population in 2018. There is no known treatment or vaccine for the highly contagious and deadly virus that causes it. It's also very hard to stop its spread, because, like SARS-CoV-2, it has a long incubation period and infected animals don't always show symptoms. It was particularly bad in China, in part because, paradoxically, the Chinese government enacted some otherwise positive regulations to curb pollution. After those regulations went into effect, industrial pig farmers couldn't upgrade their facilities fast enough, which led to farm closures and changes to China's pork supply chain. This meant sick pigs were shipped all around the country, which helped the disease spread. Initially, the government denied and then downplayed the severity of the disease. (Sound familiar?) Meat industry analysts estimate that the disease took a quarter of the world's hogs off the market and forced half of China's hog herd to be executed. The virus is particularly damaging to pig farming communities, where local economies depend on healthy herds. Today, scientists in China are developing "super-pigs," which not only resist the virus but are stronger and mature more quickly than other pigs.[36] Reportedly, they are also fortified with a gene that regulates body heat, which enables them to stay outdoors during northern China's extreme winters.[37]

In nineteenth-century Belgium, an unusual herd of cattle was discovered by local farmers. These animals were much larger than typical cattle, with an extreme amount of muscle that protruded, Schwarzenegger-style, from their backs, shoulders, loins, and rumps. They became known as Belgian Blues, and eventually, scientists discovered what made them so unusual: these cows were born with an extra gene that suppresses the production of myostatin, a protein that normally inhibits muscle growth after an animal reaches maturity. Some cows were born with two copies of that gene, leaving their frames even more engorged with muscle.[38] Belgian Blues were selectively bred to produce more meat for consumption, but researchers can now edit myostatin to improve genotypes for other mammals, including pigs, horses, goats, rabbits, and dogs. In China, myostatin was used to create double-muscled dogs intended for police work.[39]

Elsewhere in the food chain, researchers are harnessing synthetic biology to produce better feed for animals. The startup KnipBio engineers fish feed from a microbe found on leaves, editing its genome to increase carotenoids important to fish health and using fermentation to stimulate its growth. The microbes are then pasteurized, dried, and milled.

Other agricultural projects underway include synthetic organisms that can produce vast quantities of vegetable oil, and nut trees that can grow indoors using a fraction of the water these thirsty trees normally require, while also producing twice as many nuts. CRISPR has increased the level of omega-3s in plants and aided the creation of non-browning apples, drought-resistant rice, and mushrooms that can withstand jostling during transportation. (In a nod to some consumers' sentiments, in most countries product labels identify such products as genetically modified.)

Many countries lack the land mass, climate, or infrastructure to grow high-quality produce. In another era—the 1840s—a naturally occurring blight (plus the British government's terrible policies) led to the potato famine that decimated Ireland, and similar vulnerabilities still exist worldwide. Now, though, scientists and farmers can bring traditional agriculture indoors and underground, where they can use high-tech robotics, irrigation, and lighting systems to cultivate food. With these new methods, the progress of every single crop can be quantified—by using sensors, algorithms, and optimization analytics—right down to a single cherry tomato hanging on a particular vine. The expense of robots, artificial lighting, and other equipment once made indoor farming difficult to scale. But that's changing as the ecosystem matures and technology improves.

Vertical farming projects are now scattered across the globe, mostly in urban centers such as Berlin and Chicago. But Japan leads the world when it comes to indoor farming. The government has subsidized many of these operations, but they thrive thanks to Japanese consumer demand for fresh, local, pesticide-free food.[40] The Kansai Science City Microfarm, near Kyoto, uses artificial intelligence and collaborative robots to raise seedlings, replant, water, adjust lighting, and harvest fresh produce. Using complex algorithms and sensors attached to plants, the researchers track a tremendous amount of data—from carbon dioxide and temperature to water levels and plant tissue health, constantly analyzing the best conditions and systems for growing the most nutritious, tastiest foods possible. Also near Kyoto, the Kameoka-based company Spread uses machines and robots to produce between twenty thousand and thirty thousand heads of lettuce per day. Such plants mature twice as fast as lettuce grown outdoors (in around forty days), at which point they're delivered to nearby supermarkets in Japan.

Microsoft operates FarmBeats, a sort of Internet of Things for farms, on its Azure Marketplace. The company is testing the technology on two US farms as part of a multiyear plan to modernize agriculture with data analytics. The system uses unlicensed, long-range TV frequencies to connect with and capture data from solar-powered sensors, while drones gather aerial footage of crops. Machine learning algorithms mine and refine the data before sending analyses back to farmers with recommendations on which variables to tweak.

Which is all to say that, by 2030, you could be shopping at a grocery store full of fresh, nutritious, CRISPR-edited foods. What you find there will be grown nearby: underneath the store itself, perhaps, or in an adjacent vertical farm. Or even in a meat laboratory in your town.

### *The End of Meat*

It's plausible that by the year 2040, many societies will think it's immoral to eat traditionally produced meat and dairy products. Some luminaries have long believed this was inevitable. In his essay "Fifty Years Hence," published in 1931, Winston Churchill argued, "We shall escape the absurdity of growing a whole chicken in order to eat the breast or wing, by growing these parts separately under a suitable medium."[41]

That theory was tested in 2013, when the first lab-grown hamburger made its debut. It was grown from bovine stem cells in the lab of Dutch stem cell researcher Mark Post at Maastricht University, thanks to funding from Google cofounder Sergey Brin. It was fortuitous that a billionaire funded the project, because the price to produce a single patty was $375,000.[42] But by 2015, the cost to produce a lab-grown hamburger had plummeted to $11.[43] Late in 2020, Singapore approved a local competitor to the slaughterhouse: a bioreactor, a high-tech vat for growing organisms, run by US-based Eat Just, which produces cultured chicken nuggets. In Eat Just's bioreactors, cells taken from live chickens are mixed with a plant-based serum and grown into an edible product.[44] Chicken nuggets produced this way are already being sold in Singapore, a highly regulated country that's also one of the world's most important innovation hotspots. And the rising popularity of the product could accelerate its market entry in other countries.

An Israel-based company, Supermeat, has developed what it calls a "crispy cultured chicken," while Finless Foods, based in California, is developing cultured bluefin tuna meat, from the sought-after species now threatened by long-standing overfishing. Other companies, including Mosa Meat (in the Netherlands), Upside Foods (in California, formerly known as Memphis Meats), and Aleph Farms (in Israel), are developing textured meats, such as steaks, that are cultivated in factory-scale labs. Unlike the existing plant-based protein meat alternatives developed by Beyond Meat and Impossible Foods, cell-based meat cultivation results in muscle tissue that is, molecularly, beef or pork.

Two other California companies are also offering innovative products: Clara Foods serves creamy, lab-grown eggs, fish that never swam in water, and cow's milk brewed from yeast. Perfect Day makes lab-grown "dairy" products—yogurt, cheese, and ice cream. And a nonprofit grassroots project, Real Vegan Cheese, which began as part of the iGEM competition in 2014, is also based in California. This is an open-source, DIY cheese derived from caseins (the proteins in milk) rather than harvested from animals. Casein genes are added to yeast and other microflora to produce proteins, which are purified and transformed using plant-based fats and sugars. Investors in cultured meat and dairy products include the likes of Bill Gates and Richard Branson, as well as Cargill and Tyson, two of the world's largest conventional meat producers.

Lab-grown meat remains expensive today, but the costs are expected to continue to drop as the technology matures. Until they do, some companies are creating hybrid animal-plant proteins. Startups in the United Kingdom are developing blended pork products, including bacon created from 70 percent cultured pork cells mixed with plant proteins. Even Kentucky Fried Chicken is exploring the feasibility of selling hybrid chicken nuggets, which would consist of 20 percent cultured chicken cells and 80 percent plants.

Shifting away from traditional farming would deliver an enormous positive environmental impact. Scientists at the University of Oxford and the University of Amsterdam estimated that cultured meat would require between 35 and 60 percent less energy, occupy 98 percent less land, and produce 80 to 95 percent fewer greenhouse gases than conventional animals farmed for consumption.[45] A synthetic-biology-centered agriculture also promises to shrink the distance between essential operators in the supply chain. In the future, large bioreactors will be situated just outside major cities, where they will produce the cultured meat required by institutions such as schools, government buildings and hospitals, and perhaps even local restaurants and grocery stores. Rather than shipping tuna from the ocean to the Midwest, which requires a complicated, energy-intensive cold chain, fish could instead be cultured in any landlocked state. Imagine the world's most delicate, delicious bluefin tuna sushi sourced not from the waters near Japan, but from a bioreactor in Hastings, Nebraska.

Synthetic biology will also improve the safety of the global food supply. Every year, roughly 600 million people become ill from contaminated food, according to World Health Organization estimates, and 400,000 die.[46] Romaine lettuce contaminated with *E. coli* infected 167 people across 27 states in January 2020, resulting in 85 hospitalizations.[47] In 2018, an intestinal parasite known as Cyclospora, which causes what is best described as explosive diarrhea, resulted in McDonald's, Trader Joe's, Kroger, and Walgreens removing foods from their shelves. Vertical farming can minimize these problems. But synthetic biology can help in a different way, too: Often, tracing the source of tainted food is difficult, and the detective work can take weeks. But a researcher at Harvard University has pioneered the use of genetic barcodes that can be affixed to food products before they enter the supply chain, making them traceable when problems arise.

That researcher's team engineered strains of bacteria and yeast with unique biological barcodes embedded in spores. Such spores are inert, durable, and harmless to humans, and they can be sprayed onto a wide variety of surfaces, including meat and produce. The spores are still detectable months later even after being subjected to wind, rain, boiling, deep frying, and microwaving. (Many farmers, including organic farmers, already spray their crops with *Bacillus thuringiensis* spores to kill pests, which means there's a good chance you've already ingested some.) These barcodes could not only aid in contact tracing, but be used to reduce food fraud and mislabeling.[48] In the mid-2010s, there was a rash of fake extra virgin olive oil on the market. The Functional Materials Laboratory at ETH Zurich, a public research university in Switzerland, developed a solution similar to the one devised at Harvard: DNA barcodes that revealed the producer and other key data about the oil.

## HEALTHIER PLANET

The raw materials required of modern society—fuel, fibers, and chemicals—consume extraordinary resources while contributing to environmental waste and carbon dioxide emissions. Until recently, there were no alternatives: cars and trucks relied on petroleum; fashion needed traditionally grown cotton, leather harvested from cows, and massive amounts of water for production; and cutting greenhouse gas emissions meant regulating industries. The bioeconomy will bring engineered alternatives to raw materials and novel solutions to our growing CO2 problem.

### *Biofuels*

After the fall of Amyris (see Chapter 3), some doubted whether biofuels should be an ambition of synthetic biology. While sea-growing algae have been studied since the 1970 oil shocks as a less geopolitically risky alternative to fuel, the petroleum industry hasn't been supportive. Chevron, Shell, and BP each dedicated modest resources to algae biofuel research between 2009 and 2016, but most of those programs are now defunct. A small group of researchers at ExxonMobil is still working on gene editing and algae, but in 2013, the company's then CEO, Rex Tillerson, acknowledged that biofuels were three decades away from commercial viability.[49] But the hurdle for biofuels isn't just technology, it's market resistance. The traditional petroleum players aren't eager to change their core business practices, and without companies to help build the ecosystem, a future biofuel product has very little chance of commercial success. Within governments, however, research projects are still underway: the US Department of Energy awarded the J. Craig Venter Institute a $10.7 million, five-year grant to develop biofuels, and the Bioenergy Technologies Office, a group within the Department of Energy, runs an R&D program to explore algae as a possible source of fuel.[50,51] Even as the automotive industry shifts toward electric vehicles, what we learn from current biofuel projects could be applied to other sectors, such as airlines.

### *Greener Fashion*

The textile and clothing industry is a notorious polluter, but the fashion industry is working to make its practices more sustainable. Transforming cotton into fibers and textiles for clothing still relies on coal, however, and the process contributes 10 percent of global carbon emissions. Producing clothing requires a tremendous amount of water, and washing clothes made from polyester releases 500,000 tons of microfibers into the oceans each year. That's the equivalent of 50 billion plastic bottles. Roughly 85 percent of textiles ends up in landfills every year. These are clothes that have gone unsold in stores, and have to be discarded to make way for the new season's replacements, or that people simply throw away when they don't want them anymore. That's enough to fill Sydney Harbor, the biggest and deepest natural harbor in the world—*every year*.[52]

But consider if microfibers could instead be grown in a biofoundry. Bolt Threads developed a synthetic "microsilk" fabric, engineered from spider DNA, that Stella McCartney used in her line for a 2017 fashion show. A Japanese startup, Spiber, synthesized enough fibers to manufacture a limited-edition parka. Synthetic biology processes can transform mycelium—the fuzzy, fibrous structures that help fungi grow—into rugged material resembling leather. Whereas it takes years for a cow to mature to slaughter for its hide—during which time that cow must be fed, housed, and cared for—it takes just a few weeks for a spore to grow into mycelium leather. Hermès, famous for its highly coveted leather handbags, partnered with startup MycoWorks in 2021 to develop sustainable textiles made out of mycelium.[53] If fibers are designed and grown, rather than harvested and processed, then there are other opportunities on the horizon: bio-based pigments used to dye textiles could be edited to deposit the optimal amount of color, for example, with less water (or none at all) and be fully biodegradable.

Consider what synthetic biology could do for the nylon industry. Nylon is cheap to produce and durable, so it shows up everywhere: running shoes, rubber tires, cookware, camping tents, luggage, flak jackets, backpacks, tennis rackets, and more. Its production generates more than 60 million tons of greenhouse gases annually. But it's now possible to produce nylon using engineered microorganisms. Two startups, Aquafil and Genomatica, are doing just that.[54]

### *Unbreakable Everything*

Several companies are developing bio-based, ultra-durable hard biofilms and coatings so that chipped nails, scratched paint, and cracked screens become yesterday's problem. Zymergen developed a transparent biofilm that is thin, flexible, and durable enough to be used to transmit touch on a variety of surfaces, including smartphones, TV screens, and skin. Other possible applications include nearly invisible printed electronics that flex and move as needed. Imagine a football covered in a biofilm that could reveal, in real time, the ball's spin rate and velocity, along with the quarterback's precise hand placement.

But when used for screens and wearables, biofilms wouldn't just replace current surfaces; they would radically alter how we design them: instead of flat or even folding phones, we might instead see rollable screens. Picture a device the size and shape of a mechanical pencil: click the top, and you release not a pencil nib, but a retractable screen. Once expanded, the screen would snap into position and allow you to read a book or the latest news, or watch a movie. When you were finished, you'd click to retract the screen, slip the device into your pocket or purse, and go about your day.

Synthetic biology also points to a future in which our packaging and shipping materials could be far more sustainable than they are today. The insides of soda cans could be coated with a fully biodegradable film rather than plastic, as they are now. New bio-packaging could be designed to withstand heat or cold, revolutionizing the logistically complex, energy-intensive, environmentally damaging cold chain we use today to transport perishables. Even batteries in the distant future could look quite different than they do today. If we've figured out how to fatten up solar-fed bacteria, why not develop fields of artificial, leafy bio-machine plants to feed on sugar and create energy as a byproduct? Rather than buying, and ultimately throwing away, traditional batteries, which degrade and leak mercury, lead, cadmium, and other harmful metals into the environment, biological batteries could be an abundant source of clean energy.

### *Biosequestration*

Carbon dioxide is the undisputed culprit behind climate change. What if we could just suck it out of the air? Trees do that naturally, but after years of deforestation, there simply aren't enough of them to make a sizable impact on all the CO2 that we have been pumping into the atmosphere. Scientists at Columbia University are developing plastic trees that passively soak up carbon dioxide from the air and store it on a honeycomb-shaped "leaf" made of sodium carbonate—baking soda. So far, these fake trees are proving to be a thousand times more efficient at soaking up CO2 than real trees. Trees can take decades to grow, but perennials with larger leaves—hostas, elephant ears, cannas—mature quickly and propagate easily. Genetically engineering those shrubs and ground coverings, which are commonly used in residential landscapes, would dent atmospheric carbon concentrations.

The next challenge will be to purify carbon dioxide so it can be used in other processes, or to bury it safely beneath the ocean floor. One approach is to convert atmospheric CO2 into carbon nanofibers that can be used for consumer and industrial products, including wind turbine blades or airplanes. Another option comes from chemists at George Washington University, who are experimenting with what they call "diamonds from the sky." Those scientists bathe carbon dioxide in molten carbonates at 750°C (1,380°F), then introduce atmospheric air and an electrical current on nickel and steel electrodes. The carbon dioxide dissolves, and carbon nanofibers—the diamonds—form on the steel electrode. CO2 could be converted into other usable materials. The startup Blue Planet developed a way to convert it into a synthetic limestone that can be used as an industrial coating or mixed in with concrete. The company's bicarbonate rocks were included in the reconstruction of San Francisco International Airport.

These advances in materials couldn't come at a better time. The notorious pile of trash floating in the Pacific Ocean is actually two distinct collections of garbage, collectively known as the Pacific Trash Vortex. In

2018, researchers found that it is sixteen times larger than original estimates, at least three times the size of France, or a total of 617,763 square miles.[55] An estimated 5 trillion pieces of plastic float in the ocean, an amount so large that environmentalists called on the United Nations to declare the garbage patch its own country, "The Trash Isles."[56] A report by the British government warned that if we do not address the problem, the amount of plastic in the ocean could triple by 2050. The attention to the vortex prompted some innovative approaches to cleaning it up, however. One team of researchers is isolating and synthesizing the gelatinous mucus made by jellyfish in the hope of using it as a trapping agent for microplastics; it could also be used as a filter in wastewater treatment plants, or to filter the wastewater from industrial processes. In the future, plastic-eating enzymes could break down larger pieces of plastic to aid in recycling efforts.

Similarly, specialized microbes could digest polymers from unused textiles—or your old, worn-out jeans—to convert them into new fibers, spun into new textiles, and turned into new clothing. Other microbes could be designed to transform industrial wastewater, agricultural runoff, and even sewage, turning it all back into clean water.

These benefits of synthetic biology, whether close at hand or in a more distant future, illustrate how life as we know it could change. We could have personalized health care, a solution to our planet's growing food insecurity crisis, safer approaches to industrial manufacturing and agriculture, new ways to address our climate emergency, even a realistic path to off-planet living. But these futures also raise serious questions about equity, ethical challenges, geopolitical risk, and future threats to national security. The downstream implications of manipulating life are profound. Synthetic biology will influence our societies, economies, national security, and geopolitical alliances in almost inconceivable ways, which we'll detail in the next chapter.

| SEVEN |

# NINE RISKS

ANYONE WHO HAS COOKED WITH MUSHROOMS, AND ESPECIALLY THE white button mushrooms commonly used in omelets, pizzas, and spaghetti sauce, knows that they start turning brown just after they're cut. That happens because of the oxidation that occurs when the mushroom is exposed to air, and specifically, because of a gene that codes for an enzyme called polyphenol oxidase. But in 2015, Yinong Yang, a scientist at Pennsylvania State University, used CRISPR to edit six mushroom genes, which reduced that enzyme's activity by 30 percent. The result: mushrooms that stayed white longer in the package, didn't brown as easily when sliced, and could withstand being handled by automated harvesting robots.[1]

After making his discovery, Yang followed the established protocol: he sent a letter to the US Department of Agriculture explaining the methods he'd used. Because he had merely edited the mushroom's existing genome, rather than introducing foreign DNA sequences from other plants into it, Yang contended that his anti-browning mushroom should be exempt from regulation.[2] Earlier GMOs, like Monsanto's Roundup Ready soybeans, used foreign genes to engineer plants that could survive a lethal weed killer. In Yang's case, he'd merely turned an enzyme off. The edit didn't pose a danger to humans, and it was unlikely that the edits would impact other plants or animals if the mushrooms accidentally ended up in the wild. As far as biological discoveries go, this one was brilliant, both elegantly simple and incredibly boring.

Still, the news quickly leaked, and a frenzied public debate erupted over potentially dangerous "frankenfungi" and the future of genetically modified foods. "Here come the unregulated GMOs," began a story in the *MIT Technology Review*. "People are arguing about whether genetically modified foods should carry labels. But the next generation of GMOs might not only be unlabeled—they might be unregulated."[3] *Scientific American* published a lengthy article with an ominous title: "Gene-Edited CRISPR Mushroom Escapes US Regulation."[4] Dozens of non-science media outlets, including the UK's *The Independent*, *Sina* (China), and—go figure—*The Weather Channel*, stoked anxieties with provocative stories about the dangers of unregulated, edited mushrooms.[5,6,7] Fearing backlash among consumers, the Pennsylvania-based Giorgio Mushroom Company, which had helped pay for Yang's research, abruptly backtracked, insisting that it had never intended to bring CRISPR mushrooms to market.

The "mushroom problem," as it became known among synthetic biology researchers, sprang up because consumers, the media, and the regulatory system were not at all prepared for such a development. In the United States, biotech regulations have been a mess since the 1990s, when Monsanto first introduced its GMO crops. Back then, the existing regulatory frameworks had been built for traditional farming and didn't cover genetically engineered plants. Monsanto spent millions of dollars on lobbying and public relations, which forced regulators to rush. Rather than developing a modern framework to keep pace with the evolution of biology, policy makers instead cobbled together existing rules into a patchwork regulatory system. Very little has changed since. In April 2018, the USDA announced that it would not regulate crops that have been genetically edited.[8] This announcement garnered little media attention compared to the CRISPR mushroom. But it did lead to wheat engineered to have higher fiber, soybeans with more beneficial fatty acids, and tomato plants that require less water and sunlight while producing greater yields.

The firestorm created by the mushroom was nothing compared to an announcement made just a few months later, in November 2018, when a Chinese scientist named He Jiankui clutched his tan briefcase and strode onstage at a human genome editing conference held at the University of Hong Kong. There, he told a crowd of scientists packed into the auditorium that he had edited human embryos using CRISPR, supposedly to give them lifetime immunity from HIV infection. He'd performed an experiment, he said, to mimic a mutation called CCR5Δ32, which naturally occurs in some northern Europeans. This mutation resulted in thirty-two base pairs of the CCR5 protein to be deleted; this edit, He said, would prevent the AIDS-causing HIV virus from infecting an important class of cells in the human immune system.[9]

He had been working for several years on the experiment, starting with mice and advancing to monkeys before recruiting eight couples, whose sperm and eggs had been collected to create embryos, which were then edited. He stated that he had consulted with scientists in China, the United States, and Europe, and that he had submitted his research for publication in peer-reviewed journals. He also insisted that he'd obtained the parents' consent, that they were fully informed, had signed documents proving as such, agreed to implantation, and—curiously—declined to have amniocentesis to check for genetic abnormalities during their pregnancies. At this point, his peers in the audience were visibly alarmed at what they were hearing. But He plowed on, saying he had worked hard to minimize unintended effects during his experiments, such as accidentally altering any additional genes. "I feel proud," He said. And then, his big reveal: the edited embryos had resulted in a successful pregnancy with twins. His genetically edited babies, code-named Lulu and Nana, had been born just a few weeks earlier and were living in China under supervised care.[10]

Once He's announcement made its way from an academic conference to news outlets worldwide, researchers pored over his work, dissecting his self-reported methodology and outcomes. One early finding was that the CCR5 protein edits He had made wouldn't necessarily result in HIV immunity. That's because most HIV infections originate when the virus attaches itself to a different protein—CD4. Once that happens, the virus then must attach itself to a second protein, which is sometimes CCR5, but sometimes another protein. And while some strains of HIV require CCR5 to fuse to a cell and inject its code, many other variants do not.[11]

For that reason, researchers determined, Lulu and Nana likely didn't have lifetime HIV immunity. However, He's experiment probably resulted in genetic changes to their brains. In 2016, a team of researchers at the Western University of Health Sciences and the University of California at Los Angeles found that editing CCR5 significantly improved cognitive ability and memory in mice.[12] Their research was published in peer-reviewed journals and inspired many other studies. Did it also inspire He to perform cognitive genetic experiments under the guise of HIV prevention? The experiment might have enhanced the girls' ability to learn and form memories, as it has in mice. In other words, it might have made them smarter.

He's peers in the auditorium and in the broader scientific community, as well as bioethicists and politicians around the world, immediately condemned his experiment. He had violated the global consensus against human germline editing to make genetic alterations that would be permanent, and even heritable in any offspring. While He offered to produce signed consent forms, it seemed implausible that the couples who underwent IVF had been adequately informed of the experiment and its risks, and that, given the gravity of its implications, they gave their blessing. No other scientists came forward to say that they'd been in contact with He during the various phases of his research, or that they had reviewed his initial findings. He was slated to speak at another session at the conference in Hong Kong, but the organizers took him off the schedule. The title of that second panel? "The Roadmap Towards Developing Standards for Safety and Efficacy for Human Germline Gene Editing and Moral Principles."[13]

But here's the thing: No regulation exists explicitly outlawing the deliberate alteration of human embryos. The Chinese Communist Party (CCP) officially permitted gene-editing experiments on embryos in 2003, as long as the embryos were only viable for fourteen days. If what He said was true, and his experiment resulted in live births, he had violated that regulation. Regardless, the CCP now found itself embroiled in a global uproar. Ever sensitive to the optics, it began censoring all mentions of He and the edited twins on Chinese social media channels. He had embarrassed the government, revealed an ineffective regulatory system, and made it brutally clear that ethical pacts between scientists lacked any enforcement mechanism. His announcement had also led to wild speculation that he had blown the whistle on a CCP-sponsored eugenics program designed to create super-intelligent Chinese people in order to increase its competitiveness against the United States. In 2020, a Chinese court sentenced him to three years in prison for "illegal medical practice," and it handed down shorter sentences to two colleagues who had assisted him.[14]

The mushroom problem and the CRISPR-edited twins illustrate the spectrum of risks associated with synthetic biology. The twins represent a worrisome new reality in which highly skilled specialists could unilaterally make decisions that affect the future of humanity. Anti-browning mushrooms won't harm the environment if planted in nature, though they do reveal the shortcomings of our current regulatory frameworks and the general public's lack of basic, biological literacy. But what's really worrisome are the future risks not being openly discussed. The same deleted genes that keep mushrooms white could be enhanced and reintroduced to create mushrooms that brown and rot quickly. Radical antiglobalization activists could engineer our produce so that it spoiled quickly and wouldn't be able to withstand transport. This would decimate the world's food supply, much of which today relies on trade between countries. And where do we draw the line between improvement and enhancement? If we choose to edit genomes to improve health outcomes, those changes could include resistance to obesity and improved muscle function. Would society be cleaved into genetically enhanced people and those who had to accept their inherited fates? These questions are important beyond the obvious ethical and philosophical concerns. Once it's unleashed, biology self-replicates at scale, reproducing and sustaining itself for generations. Synthetic biology is often permanent. As the nine risks we'll discuss show, unless we enter this new age wisely, we may lose control over what comes next.

## RISK #1: DUAL USE IS INEVITABLE

In 1770, German chemist Carl Wilhelm Scheele performed an experiment and noticed he'd created a noxious gas. He named it "dephlogisticated muriatic acid." We know it today as chlorine.[15] Two centuries later, a German chemist, Fritz Haber, invented a process to synthesize and mass-produce ammonia, which revolutionized agriculture by generating the modern fertilizer industry. He won the Nobel Prize in Chemistry in 1918. But that same research, combined with Scheele's earlier discovery, helped create the chemical weapons program that Germany used in World War I.[16]

This is an example of the "dual use dilemma," in which scientific and technological research is intended for good, but can also, either intentionally or accidentally, be used for harm. In both chemistry and physics, the dual-use dilemma has long been a concern, and it has led to international treaties limiting the most worrisome applications of problematic research. Because of the Convention on the Prohibition of the Development, Production, Stockpiling and Use of Chemical Weapons and on Their Destruction (otherwise known as the Chemical Weapons Convention, or CWC), a treaty signed by 130 countries, many dangerous chemicals that are sometimes used in scientific or medical research have to be monitored and inspected. One example is ricin, which is produced naturally in castor seeds and is lethal to humans in the tiniest amounts. A brief exposure in a mist or a few grains of powder can be fatal, so it is on the CWC list. Triethanolamine, which is used to treat ear infections and impacted earwax, and is an ingredient to thicken face creams and balance the pH of shaving foams, is also listed, because it can also be used to manufacture HN3, otherwise known as mustard gas.

Similar international treaties, enforcement protocols, and agencies exist to monitor dual uses in chemistry, physics, and artificial intelligence. But synthetic biology is so new that such treaties don't yet exist for it, even though, within the scientific community, there have been decades of discussions about how to prevent harm.

A team of researchers at the State University of New York at Stony Brook conducted an experiment in 2000–2002 to determine whether they could synthesize a live virus from scratch using only publicly available genetic information, off-the-shelf chemicals, and mail-order DNA. (The project was financed with $300,000 from DARPA, as part of a program to develop biowarfare countermeasures.) The researchers purchased short stretches of DNA and painstakingly pieced them together, using nineteen additional markers to distinguish their synthetic virus from the natural strain they were attempting to reproduce.

They succeeded. On July 12, 2002—just after Americans celebrated their first Fourth of July after the 9/11 terrorist attacks, after which jittery millions were relieved that another horrific event didn't happen on that day—those scientists announced that they had re-created the poliovirus in their lab using code, material, and equipment that anyone, even Al-Qaeda, could get their hands on. They'd made the virus to send a warning that terrorists might be making biological weapons, and that bad actors no longer needed a live virus to weaponize a dangerous pathogen like smallpox or Ebola.[17]

Poliovirus is perhaps the most studied virus of all time, and at the time of the experiment there were samples of the virus stored in labs around the world. The goal of this team's work wasn't to reintroduce poliovirus into the wild, but to learn how to synthesize viruses. It was the first time anyone had created this type of virus from scratch, and the Department of Defense hailed the team's research as a massive technical achievement. Knowing how to synthesize viral DNA could help the United States gain new insights into how viruses mutate, how they become immune to vaccines, and how they could be developed as weapons. And while creating a virus to study how it might be used as a bioweapon may sound legally questionable, the project didn't violate any existing dual-use treaties. Not even a 1972 treaty explicitly banning germ weapons, which outlaws manufacturing disease-producing agents—such as bacteria, viruses, and biological toxins—that could be used to harm people, animals, or plants. Nonetheless, the scientific community was incensed. Intentionally making a "synthetic human pathogen" was "irresponsible," Craig Venter said at the time. But this was no isolated incident.[18]

The World Health Organization declared smallpox eradicated in 1979. It marked a major human achievement, because smallpox is a truly diabolical disease—extremely contagious, and with no known cure. It causes high fever, vomiting, severe stomachache, a red rash, and painful, yellowish, pus-filled domes all over the body, which start inside the throat, then spread to the mouth, cheeks, eyes, and forehead. As the virus tightens its grip, the rash spreads: to the soles of the feet, the palms of the hands, the crease in the buttocks, and all around the victim's backside. Any movement pressures those lesions until they burst through nerves and skin, leaving behind a trail of thick fluid made of flaky, dead tissue and virus.

Only two known samples of natural smallpox exist: one is housed at the CDC, the other at the State Research Center of Virology and Biotechnology in Russia. For years, security experts and scientists have debated whether to destroy those samples, because no one wants another global smallpox pandemic. That debate was made moot in 2018, when a research team at the University of Alberta in Canada synthesized horsepox, a previously extinct cousin of smallpox, in just six months, with DNA they'd ordered online. The protocol for making horsepox would also work for smallpox.[19]

The team published an in-depth explanation of how they synthesized the virus in *PLOS One*, a peer-reviewed, open-access scientific journal that anyone can read online. Their paper included the methodology they used to resurrect horsepox along with best practices for those who wanted to repeat the experiment in their own labs. To the team's credit, before publishing its research, its lead investigator followed protocol, just like Yang did with his mushroom, and alerted the Canadian government. The team also disclosed its competing interests: one of the investigators was also the CEO and chairman of a company called Tonix Pharmaceuticals, a biotech company investigating novel approaches to neurological disorders; the company and the university had filed a US patent application for "synthetic chimeric poxviruses" a year earlier. No one—not the Canadian government, nor the journal's editors—sent back a request for them to rescind the paper.

The poliovirus and horsepox experiments dealt with synthesizing viruses using technology designed for well-intentioned purposes. What scientists and security experts fear is different: terrorists not only synthesizing a deadly pathogen, but intentionally mutating it so that it gains strength, resilience, and speed. Scientists conduct such research in high-security containment labs, attempting to anticipate worst-case-scenario pathogens by creating and studying them.

Ron Fouchier, a virologist at the Erasmus Medical Center in Rotterdam, announced in 2011 that he'd successfully augmented the H5N1 bird flu virus so that it could be transmitted from birds to humans, and then between people, as a new strain of deadly flu. Before COVID-19, the H5N1 virus was the worst to hit our planet since the 1918 Spanish flu. At the time that Fouchier conducted his experiment, only 565 people were known to have been infected with H5N1, but it had a high mortality rate: 59 percent of those who'd been infected died. Fouchier, then, had taken one of the most dangerous naturally occurring flu viruses we had ever encountered and made it even more lethal. He told fellow scientists that he'd "mutated the hell" out of H5N1 to make it airborne, and therefore significantly more contagious. There was no H5N1 vaccine. The existing virus was already resistant to the antivirals approved for treatment. Fouchier's discovery, which was funded in part by the US government, scared scientists and security experts so much that, in an unprecedented move, the National Science Advisory Board for Biosecurity, within the National Institutes of Health, asked the journals *Science* and *Nature* to redact parts of his paper ahead of publication. They feared that some of the details and mutation data could enable a rogue scientist, hostile government, or group of terrorists to make their own, hyper-contagious version of H5N1.[20]

We've just lived through a global pandemic that no one wants to see replicated. We may have COVID-19 vaccines, but we're still coexisting with the virus. As of this writing, there are several concerning variants in the United States, which include strains from the United Kingdom (B.1.1.7), South Africa (B.1.351), Brazil (P.1) and India (B.617.2, known as the Delta variant). Before we eradicate SARS-CoV-2, as we eventually did with smallpox, there will be more mutations and many new strains. Some could affect the body in ways we've not yet seen, or even imagined. But there is tremendous uncertainty over how, and when, the virus could mutate.

Obviously, one would hope that virus research would be undertaken in a lab where fanatical adherence to safety and rigorous oversight policies were strictly enforced. Just before the World Health Organization declared smallpox eradicated, a photographer named Janet Parker was working at a medical school in Birmingham, England. She developed a fever and body aches, and a few days later, a red rash. At the time, she thought it was chicken pox. (That vaccine had not yet been developed.) The tiny, pimple-like dots she'd been expecting, however, developed into much bigger lesions, and they were full of a yellowish, milky fluid. As her condition worsened, doctors determined that she'd contracted smallpox, almost certainly from a sloppily managed high-security research lab inside the same building where she worked. The lab's head researcher committed suicide just after Parker was diagnosed. Parker, sadly, is now remembered as the last person known to have died from smallpox.[21]

Does the benefit of being able to accurately predict virus mutations outweigh the public risks of gain-of-function research (that is, research that involves intentionally mutating viruses to make them stronger, more transmissible, and more dangerous)? It depends on who you ask. Or, rather, which agency you ask. The National Institutes of Health issued a series of biosafety guidelines for research on H5N1 and other flu viruses in 2013, but the guidelines were narrow and didn't cover other kinds of viruses. The White House Office of Science and Technology Policy announced a new process to assess the risks and benefits of gain-of-function experiments in 2014. It included influenza along with the MERS and SARS viruses. But that new policy also halted existing studies intended to develop flu vaccines. So the government reversed course in 2017, when the National Science Advisory Board for Biosecurity determined that such research wouldn't pose a risk to public safety. In 2019, the US government said that it had resumed funding for—wait for it—a new round of gain-of-function experiments intended to make the H5N1 bird flu more transmissible again. Meanwhile, this back and forth doesn't stop bad actors from gaining access to open-source research papers and mail-order genetic material.

When it comes to synthetic biology, security experts are particularly concerned about future dual-use issues. Traditional force protection—the security strategies to keep populations safe—won't work against an adversary that has adapted gene products or designer molecules to use as bioweapons. Dr. Ken Wickiser, a biochemist and associate dean of research at the military academy West Point, in an August 2020 paper published in the academic journal *CTC Sentinel*, which focuses on contemporary terrorism threats, wrote: "As molecular engineering techniques of the synthetic biologists become more robust and widespread, the probability of encountering one or more of these threats is approaching certainty. . . . The change to the threat landscape created by these techniques is rivaled only by the development of the atomic bomb."[22]

## RISK #2: BIOLOGY IS UNPREDICTABLE

After the Human Genome Project, Craig Venter and his team switched from reading genomes to writing them. They had a singular goal: to create an organism with the smallest genome possible that could still survive and reproduce on its own. Venter wondered: If he could edit down the microbe's genome to its minimal viable parts, could he reveal the source code for life? And if we gained that knowledge, could we construct entirely new forms of life? Venter and his collaborator Hamilton Smith hypothesized that a minimally viable genome might act as a basic chassis, the scaffolding upon which other genes could be added for new functionality. They used an organism called the *Mycoplasma genitalium*, which had a minuscule genome, to see whether they could synthesize a new version with slightly different code. They made an astonishing discovery in May 2010: they could destroy the DNA in the existing *Mycoplasma* cell and replace it with DNA they had written, and the cell would then self-replicate. They named it JCVI-syn1.0, or Synthia, for short. The synthetic bacterium that we described in Chapter 1—the one watermarked with quotes from J. Robert Oppenheimer, poems from James Joyce, and the names of researchers who worked on the project? That was Synthia.

It was, according to Venter, the first self-replicating species on the planet whose parent was a computer. Or, more precisely, whose parents were a team of twenty scientists and a cluster of computers, resulting from the thousands of selections those humans and machines had made while working together. Synthia was "a living species now, part of our planet's inventory of life," Venter said. The project was designed to help Venter's team understand the basic principles of life: a minimal cell is an analog to the last universal common ancestor connecting all life on Earth.[23]

Before announcing the team's discovery, Venter sent a message to the Obama White House asking to brief officials on a wide range of policy implications, security challenges, and ethical issues the project raised. At first, administration officials didn't know what to make of Synthia. They considered classifying the research, but many in the synthetic biology community already knew about Venter's minimal viable genome project. They recommended publishing the research, but directed the Presidential

Commission for the Study of Bioethical Issues to study the implications of this milestone and deliver a report within six months, with recommendations for what the government should do next, if anything.

Andrew was quoted in the *Ottawa Citizen* newspaper, and he said, "Venter deserves the Nobel Prize for his pioneering work in creating a 'new branch on the evolutionary tree—one where humans shape and control new species.'"[24] Not everyone shared his optimistic sentiment. Unsurprisingly, intense media coverage and wild speculation followed the announcement. "This is a step towards something much more controversial: creation of living beings with capacities and natures that could never have naturally evolved," Julian Savulescu, an ethics professor at Oxford University, told *The Guardian*. "The potential is in the far future, but real and significant: dealing with pollution, new energy sources, new forms of communication. But the risks are also unparalleled. These could be used in the future to make the most powerful bioweapons imaginable."[25] The ETC Group, an activist organization and critic of biotechnology, compared Venter's creation to the splitting of the atom: "We will all have to deal with the fallout from this alarming experiment." Religious groups, furious that Venter was playing God, wanted him arrested.[26]

That presidential commission was directed to develop a set of criteria to weigh the benefits of minimum viable genome creation against risks. What if in the future, a human-created life form like Synthia escaped from the lab? There had never been any concerns that Venter and his team were ever anything less than meticulous and keenly observant of safety protocols. But Venter wasn't the problem. Experts worried about others who might be inspired by his research. The scientific community is insanely competitive, always racing to make new discoveries, be the first to publish in peer-reviewed journals, and stake their claims at the patent office before anyone else. The races to synthesize insulin and to map the human genome show that when it comes to scientific discoveries, there is no award for second place.

Venter and his collaborator, Smith, were already thinking beyond Synthia. They hypothesized that about 100 genes could be removed from *M. genitalium* without upending its function. But they weren't sure which 100 genes to take away. They synthesized hundreds of shrunken genomes, testing different combinations with the intention of eventually inserting viable candidates into a cell. In 2016, Venter's team created what they called JCVI-syn3.0, a single-celled organism with even fewer genes—just 473—which made it the simplest life form ever known.[27] The organism acted in ways scientists hadn't predicted. It produced oddly shaped cells as it self-replicated. Scientists came to believe that they'd taken away too many genes, including those responsible for normal cell division. They remixed the code once again, and in March 2021 announced a new variant, JCVI-syn3A. It still has fewer than 500 genes, but it behaves more like a normal cell.[28]

It's worth emphasizing, again, that the probability of the strangely behaved JCVI-syn3.0 escaping from the lab and causing any harm was negligible. But biology is highly interconnected, and it tends to self-sustain, even when we don't want it to. Creating a minimal viable genome, or any other novel organism, could lead to a cascading effect and be impossible to manage in the wild. One of the presidential commission's reports described the dangers of what's called "outcrossing," which is what happens when engineered genes mix with wild populations and native species. Outcrossing could lead to new types of weeds that could kill other plants, for example, or a new pathogenic microorganism that could spread disease to insects, birds, and other animals. A lab accident, or a containment breach, could result in today's harmless laboratory bacterium becoming tomorrow's ecological catastrophe.

## RISK #3: PRIVATE DNA IS A SECURITY RISK

In December 2019, a mysterious and anonymous organization called the Earnest Project announced that it had surreptitiously collected DNA from used breakfast forks, wine glasses, and paper coffee cups used at the World Economic Forum's annual meeting in Davos. The Earnest Project launched a website and auction catalog and announced plans to sell the genetic data of many world leaders and celebrities, including then president Donald Trump, German chancellor Angela Merkel, and musician Elton John, to the highest bidders. There was no way to verify whether the DNA samples were authentic, but, more importantly, there was no US law banning the group from selling Trump's genetic data. Alaska, New

York, and Florida have outlawed stealing someone's DNA, and it's illegal to pluck someone's hair out of their head without permission. But there is no federal law prohibiting someone from taking someone's abandoned DNA and doing what they want with it.

When Trump was in office, he—like all US presidents—was assigned Secret Service agents. Among their duties was the responsibility to sweep every location he visited, collect all the trash, and safely discard it. A DNA sample culled from a napkin Trump had dirtied or a plastic fork he'd eaten with could reveal his genetic variants—such as whether he has a mutation linked to early-onset Parkinson's or Alzheimer's disease. It could also prove (or disprove) an allegation made by *New York* magazine columnist E. Jean Carroll, who accused Trump of raping her in the 1990s. (She kept the dress she was wearing during the alleged encounter, and she has said that it contains his DNA material.) If she had access to a discarded McDonald's wrapper or a used napkin from Davos, she could get his DNA sequenced herself. But with synthetic biology, this sequence could also be used to create a personalized bioweapon. Biological weapons don't need to result in widespread death or a pandemic to be effective.

DNA is tough, able to survive for millennia under the right conditions. The vast majority of VIPs don't travel with eagle-eyed minders who sweep the environs. Amy often rode with Joe Biden on Amtrak along its Northeast Corridor route before he was elected president. On that route, the first-class car—where Biden typically rode—serves breakfast, lunch, and dinner on plates, with silverware. On another Amtrak ride, Amy sat across from Supreme Court Justice Clarence Thomas. During the trip, he sneezed several times into a paper tissue, which he left behind after he departed in New York. What if someone on these trains collected his, or Biden's, DNA samples? During the early stages of US presidential elections, no security detail cleans up every single wrapper and napkin after local campaign stops. What if, in 2023, as the next presidential election gets underway and the field of candidates is large, a bad actor collected samples and sequenced all the candidates' DNA? As the field narrowed to just two candidates, that bad actor would be in a good position to sow misinformation: fabricate an affair, claim evidence of a physical altercation, call into question that candidate's ethnicity or birthplace, stoke fear about hidden genetic diseases and the candidate's ability to lead. Or tailor microbes and viruses uniquely damaging to a candidate.

This is what made some 2019 research from Duke University particularly interesting. The Duke scientists developed programmable swarmbots, specially designed bacteria engineered to burst and release proteins on command. It's a clever idea, really: those bacteria are programmed to die if they leave the swarm. That synthetic biology technique could be used as a failsafe, to stop other genetically modified organisms from escaping their designated environments. In a proof-of-concept experiment, the scientists at Duke engineered a nonpathogenic strain of *E. coli* to produce a chemical that acted like an antidote to antibiotics. As long as the *E. coli* stayed in the swarm, they were safe, even when the researchers doused them with an antibiotic. If an individual bacterium moved too far away, it would lose its protection and die immediately. But someone could design a pathogenic microbial swarmbot that would burst and release harmful chemicals.[29]

There is also the specter of personal viruses, engineered to deliver genetic code to just one person. A landmark May 2021 study was designed to restore vision to people who have a rare genetic disease that causes blindness. It showed how CRISPR could be used to edit DNA while it's still inside patients' bodies. In people with this form of blindness, a defect in the CEP290 gene slowly destroys light-sensing cells in the retina, until little healthy tissue is left. The result is deteriorating vision; eventually, the retina becomes like just a tiny porthole to the outside world (imagine the tip of a pencil lead). Because retinas are extremely complicated and fragile, doctors cannot replace them with a transplant, and extracting cells to manipulate them in a lab is also too challenging. So instead, researchers figured out how to build a beneficial virus to carry new genetic instructions with a mandate for the cells to perform CRISPR on their own inside the retina. (Viruses, as we've explained, are just containers for biological code, so they can be beneficial or detrimental.) They injected billions of copies of this virus under the retinas of a small number of people with this form of blindness, and so far, the experiment seems to be working. CRISPR has acted like a microscopic surgeon, editing the CEP290 mutation in order to produce a protein that can restore light-sensing cells, and eventually the patients' vision. This is a thrilling, groundbreaking study. But given the propensity for dual use, it is plausible that other viruses could be engineered to trigger the opposite effect: to cause mutations, rather than correcting them.[30]

In the previous chapter, we discussed pluripotent stem cells, those cells with the ability to self-replicate and become any other cell in the human body. Pluripotent cells can be easily retrieved from left-behind genetic material. Someday, they will help people become parents more easily. But what if, in the farther future, someone used these cells to engineer, say, a slow-moving infection that targets an organ like the kidney? At first, it might look like diabetes, but then it wouldn't respond to medication. Dialysis would be followed by renal failure, and ultimately death.

It's easy to imagine other such scenarios. A disgruntled ex-employee could hold the DNA of a company's board of directors for ransom. Bad actors could scrape and sequence a CEO's microbiome, and manipulate the CEO's gut with a specially designed probiotic, causing continual gastric distress. The US Securities and Exchange Commission requires public companies to disclose when the CEO of a publicly traded company contracts an illness that is significant enough that it could have an adverse impact on business. But there aren't yet tests or disclosure requirements for biohacks.

And what about biosurveillance? The Trump administration approved a program—which, thankfully, never took effect—to collect DNA, along with a host of other biometric data, such as iris scans and palm prints, from anyone seeking to come to the United States. But the administration did begin collecting DNA samples from immigration detainees, and it houses that data in a government database. In the future, would private health insurers offer a discount in exchange for you allowing them access to your DNA? Could a life insurance company, a mortgage lender, or a bank require your DNA as part of its verification process? What if Big Tech companies—Google, Apple, Amazon—mapped your genetic data to all the other data they collect about you? Each of these companies is investing heavily in health and life sciences. Today, we talk about surveillance capitalism. Imagine if that surveillance included your genetic code.

In the future, our most worrying data security breaches could involve DNA. That means that biology, in the era we are entering, may become a major information security problem.

## RISK #4: REGULATION IS WOEFULLY BEHIND

At first glance, Josiah Zayner's faux-hawk and spikey bleached bangs, prominent piercings, and beard scruff might fool you into assuming he's the bass guitarist of a really loud punk band. He's not. He's a molecular biophysicist who holds a PhD from the University of Chicago. Zayner, who sports a tattoo that urges you to "Create Something Beautiful," for a while was a synthetic biology research fellow at NASA, where he worked on projects focusing on engineering bacteria for degradation and recycling of plastic, and hardening Martian soils. But he'd become increasingly disenchanted with space exploration. There was so much of the human body to explore.

In 2015, Zayner launched a successful crowdfunding campaign on Indiegogo to supply DIY CRISPR kits to hobbyists. In a video explaining the kits, there was a shot of petri dishes stored next to food in his refrigerator, which, to put it mildly, is an obvious breach of biosafety protocols. He raised more than $69,000, which was around seven times his original goal. He left his NASA fellowship early because of the excitement about his kits and because, as he put it, he was "fed up with the system" and with the slow pace of scientists who were just "sitting on their asses."[31] Using his Indiegogo haul, he started a new company, the Open Discovery Institute—ODIN—named for the shape-shifting Norse god of divination, magic, wisdom, and death. Why was experimentation restricted to NASA scientists with PhDs? he wondered. Shouldn't everyone be allowed—encouraged, even—to tinker with biology? Nature was already democratized; the tools to access and manipulate it should be, too.

Zayner's first post-Indiegogo project was a kit allowing anyone to genetically modify bacteria. He launched a website and began selling $160 kits for people to make glow-in-the-dark beer, by using a gene found in jellyfish. Unlike Yang and Fouchier, Zayner didn't follow any protocols. He didn't submit any of his research to peer-reviewed journals, or tell any federal regulator about his methodology. He didn't even follow the guidelines established within the DIYbio community, which, among other things, says there should be dedicated refrigerators for biomaterials—though, in fairness, we should point out there aren't any global lab standards governing biosafety.

The success of Zayner's crowdfunding campaign and its tantalizing premise—anyone can use CRISPR to edit the living world—attracted the attention of the FDA, which was none too happy that he was ignoring its rules. According to the FDA, fluorescence could be classified as a color additive; therefore, Zayner's glow-in-the-dark beer kit needed to pass through a rigorous approval process. But this was a murky regulatory area, because he wasn't selling a regulated food product, like beer. He was selling a set of genetic instructions, along with some cheap lab equipment that was perfectly legal for anyone to sell. Zayner ignored the FDA and kept on selling his kits. The FDA had no way to counter.

In the United States, regulations are a patchwork that generally covers products, not processes. The reason is simple: the government doesn't intervene until there's a problem, so as not to stifle innovation. So in the early 1970s, when scientists first discovered the genetic engineering tool of recombinant DNA, no restriction prevented researchers from swapping genes from one species into another using *E. coli* bacteria. Microbiologists heralded the achievement as a milestone. It was, but the government wasn't interested in that discovery, or what it could yield in the future.

Companies were using recombinant DNA by the 1980s to commercialize microbes and plants. There still wasn't a regulatory framework in place, though, so in 1986 the White House Office of Science and Technology Policy, which advises the president and helps to coordinate different agencies on such matters, was summoned to develop a plan. But it chose not to work through the arduous process of writing new laws to govern genetically engineered products. Instead, it retrofitted old laws under a plan called the Coordinated Framework for the Regulation of Biotechnology, which tasked three agencies—the Food and Drug Administration, the Environmental Protection Agency (EPA), and the US Department of Agriculture (USDA)—with overseeing biological advancements under the guiding principle that biotechnology is not harmful, but certain products may be. Even after an update in 1992, the framework was often ambiguous. The roles and responsibilities of each agency weren't always clear, and there was no long-term strategy to prepare those agencies for advancements in biotechnology.

Consider this: Under the Coordinated Framework the USDA regulates plants. If someone makes a microbe that's a plant pathogen, then the USDA can get involved. But if a threat to a crop isn't likely, it has no oversight. (This is what led to the mushroom problem.) The EPA's primary role is to protect human health and the environment from outside threats. This regulation excludes microorganisms used for academic research, but it does include genetically modified organisms that contain plant pest DNA, or that were made using a plant pest as a vector. That means it can regulate a biofuel, synthetic fertilizer, or pesticide if it is likely to produce toxic chemicals. If there is no such risk, the EPA has no oversight. The FDA's job is to keep our food and beverages safe, along with our medications, medical devices, and the like, and it regulates modified organisms that are used to produce drugs, food and food additives, dietary supplements, or cosmetics. So the FDA oversees all genetically modified animals to ensure that they meet safety standards for human use.

But all these regulations are difficult to enforce. It isn't like there's a USDA official standing over scientists and watching them perform research; there aren't even site inspections or routine audits. Instead, under the Coordinated Framework, companies proposing to sell things can choose to submit a voluntary review to prove that those things won't kill anyone. The mushroom didn't produce pesticides or toxic chemicals, so the EPA had no jurisdiction. The team behind it didn't use microbes to deliver DNA, so the USDA had no say in the outcome. The FDA at the time was overcommitted and underfunded, so while it could have intervened, there weren't enough resources available to deal with a novel mushroom that probably wouldn't cause harm.

The United States' patchwork approach to regulation isn't unique. The European Union, along with the United Kingdom, China, Singapore, and many other nations, approach the governance of synthetic biology in similar ways, using existing biotechnology frameworks. Which—let's face it—weren't written with JCVI-syn3.0 in mind. The United Nations convened a working group to discuss the safety implications of modified organisms that resulted in yet another framework—the Cartagena Protocol on Biosafety to the Convention on Biological Diversity. Under the protocol, countries could restrict or ban any biotechnology they deemed potentially unsafe, even if there was no evidence showing that the research posed a danger to biodiversity or security. The European Union and China were signatories, but plenty of countries, including the United States, Japan, and Russia, did not sign on, and there is no clear enforcement mechanism. The protocol only gives countries the right to bar imports of live genetically modified organisms. Countries can choose not to exercise that right, or can ask an exporting country to provide a risk assessment of the organism. But such assessments aren't performed by an independent third party. They're handled by the exporting country itself.

What if a country is found to be intentionally creating a weapon? The Biological Weapons Convention is a multilateral disarmament treaty banning the development, production, and stockpiling of bioweapons. The United States, Russia, Japan, the United Kingdom, China, and Europe all ratified the treaty. Currently, the convention applies to any form of biological weaponization, but the tricky part is evaluating harm. Let's say someone intentionally edits a weed so that it chokes off a major export crop. That would cause great economic harm to farmers, and possibly to a nation's GDP. But can you compare that to mustard gas? The convention requires countries to designate just one agency to be responsible for guaranteeing compliance with the treaty's provisions. In the United States, that agency isn't a futuristic R&D laboratory teeming with biologists. It's the Federal Bureau of Investigation. If a scientist working at a biofoundry receives a suspicious order from a customer, they're supposed to contact the FBI's Weapons of Mass Destruction Directorate. But that directorate's primary mission is to prevent WMDs, of course, and, like many federal agencies, the FBI doesn't devote substantial resources to this new area of science. It relies on researchers to police themselves.

The regulatory mess is why Zayner could launch his business. There would be very little that any of these agencies, frameworks, protocols, or directorates could do to stop him from selling DIY CRISPR kits. His fluorescent beer kit garnered international attention, and soon Zayner was known less for his genetic engineering kits than for his provocative stunts, which earned him headlines in *Bloomberg* and *The Atlantic*. Among them: he gave himself a fecal transplant—a risky procedure to manage severe gastrointestinal problems—in a hotel room using a friend's poop (after inviting a journalist from The Verge to witness the proceedings). Later, he created a DIY COVID-19 vaccine, using the code name Project McAfee, named for the antiviral software. He even developed an online course titled "Do-It-Yourself: From Scientific Paper to COVID-19 DNA Vaccine," to allow viewers to follow along with his home-brewed vaccine journey.[32]

Unsurprisingly, Zayner's embrace of his part-performance-artist, part-scientist persona netted him a fair share of criticism, but at a synthetic biology conference in 2017, he pushed the performance artist act even further. He announced to the crowd that he'd made a CRISPR cocktail that would, he proclaimed, "modify my muscle genes to give me bigger muscles!" With that, he jabbed a syringe into his forearm and told conference-goers that for $189, they could buy his DIY Human CRISPR guide along with a DIY CRISPR kit with modified DNA that would stimulate muscle growth. (The CRISPR cocktail didn't actually work.)[33]

In none of these cases has Zayner ever broken the law, even though he's clearly overstepped ethical bounds. As his home-brewed COVID-19 vaccine went live, the FDA was cracking down on unproven, untested products purported to inoculate or cure people. But he never attracted the FDA's attention. A complaint to the California Medical Board spurred it to investigate whether Zayner was practicing medicine without a license, but the investigation was dropped. The Coordinated Framework regulates damage to plants, but it doesn't regulate people who want to damage themselves. In the United States, biological self-experimentation appears to be perfectly legal, even when performed in public. German authorities tried to crack down on Zayner exporting DIY genetic engineering kits by citing a law that bans genetic engineering performed outside of licensed labs. They issued Zayner a stern warning, telling him that he could face a fine of $55,000 and up to three years in prison. However, they can't extradite him from the United States to face any penalties. As of this writing, he's made it clear on his company's website that he will still ship products to Germany, except for "perishables" such as bacteria or plasmids. Germany isn't a proxy for Europe when it comes to biohacking regulations, so if you live in the Strasbourg region of France, you could easily pick up your bacteria there, culture your cells, cross the Rhine River to Germany, and ingest (or set free) whatever you've created, without running afoul of any local regulations.[34]

Outside of a known bioweapons threat, enforcement for any of these international treaties is scant and loose. While there is an established code of ethics written by the DIY bio community's North American Congress, a citizen science association, it's not in any way legally binding. Scientists may scream, but that hasn't stopped Zayner, and it certainly didn't stop He Jiankui. All the more reason why regulators must accept that a new era in synthetic biology is already underway, and that this new approach to biology warrants a new approach to regulation.

### RISK #5: CURRENT LAWS STIFLE INNOVATION

Jennifer Doudna and Emmanuelle Charpentier published a paper in 2011 detailing how to use CRISPR to edit DNA. Doudna followed this up in 2013 with another paper showing how to use CRISPR to edit animal cells. But just a few weeks earlier, another CRISPR paper building on their earlier work was published thanks to an academic publishing loophole allowing submitters to jump to the front of the editorial queue with the payment of additional fees. Because of the loophole, Feng Zhang, a researcher at the Broad Institute (a collaboration between MIT and Harvard) technically became the first to prove, in a peer-reviewed journal, that CRISPR could be used to edit human cells. At that point, the most famous CRISPR molecule was Cas9, and that's where the battle over patents and intellectual property (IP) started.

The University of California at Berkeley and the University of Vienna, the publicly funded research centers where Doudna and Charpentier worked, respectively, filed patent applications in 2012 on CRISPR-Cas9. But the Broad Institute, a private research center, paid to fast-track its patent review for the same work. The US Patent and Trademark Office didn't switch over to a first-to-file system until March 16, 2013, so that expedited review meant that it awarded patents to Broad, and it meant that Zhang's Editas Medicine won the exclusive license to the most important patent: the one governing all future human therapeutic uses of CRISPR. UC Berkeley appealed.

Amy participated in a series of government meetings from 2016 to 2018 about genetic editing policy and oversight. She was invited in 2017 to a closed-door meeting coordinated by the State Department and the National Academies of Sciences, Engineering, and Medicine. In the room were a dozen research scientists and government officials, and the group's task was to discuss the future of CRISPR's regulation, biosecurity, and future competitiveness. Amy sat next to Zhang, who was quiet and reserved throughout the day, and while he answered questions about science, he wouldn't discuss the patent problem. By the end of the meeting, Amy had reached a troubling conclusion: the US government had no plan to manage the coming onslaught of intellectual property battles looming on the horizon. As of this writing, four years after that meeting, the Broad Institute still held the patents, which meant that anyone who wanted to use them would need to pay that organization for licenses. There are ten companies founded on the CRISPR-Cas9 patents in question.[35]

Science is iterative, and discoveries are built on the prior research of many people. Well before Doudna, Charpentier, and Zheng published their papers, in 2009, a Northwestern University postdoctoral student from Italy named Luciano Marraffini published research that showed, for the first time, that CRISPR could target DNA. Some scientists feel that with so much public money involved in discoveries such as the ones around CRISPR, no one entity should have legal ownership over its related IP. This would keep the science open and allow others to innovate on top of it, without fearing lawsuits or having to pay royalties, which can be quite expensive. Meanwhile, investors, ever aware of the legal landscape, are looking to finance new biotechnology that doesn't infringe on the existing CRISPR patents. All of this keeps becoming more complicated: as more researchers, academic institutions, and startups file their own patent applications for subtle variants to CRISPR molecules, it will be even more difficult to narrow down IP rights to just a few key players. CRISPR-Cas9 includes different enzymes that can be used to cleave DNA. Meanwhile, Cas9 isn't the only molecule that can be used to edit DNA. Cas12, Cas14, CasX, and other, less famous molecules can do the job. Those, unsurprisingly, have been patented by various organizations.

IP laws create two tangible threats. The first is obvious: patenting new discoveries is a bet on the future. When Doudna, Charpentier, and Zheng first discovered what to do with CRISPR-Cas9, there was no use case for their research, but there very likely would be someday, which would allow them to generate future revenue through commercialized products. But the second threat is more concerning: whoever receives the patent will determine how the future of research will unfold. A person, or institution, could make CRISPR-Cas molecules available to academic institutions at little or no cost. Or they could refuse to license the technology at all. It's quite plausible that by the time the lawsuits are settled, Doudna and Charpentier—who won the Nobel Prize for their work—could be prohibited from using their own discovery to advance the scientific field they may have founded.

It is becoming increasingly difficult to determine whether a Cas molecule has been patented and who owns the IP rights. Much early-stage and basic research is exploratory, so researchers or entrepreneurs would need to pay expensive licensing fees before engaging in any meaningful work—work from which profitable outcomes are by no means assured. This would mean relying on a third party, such as a publicly funded institution or government grants (read: your taxes), or on a venture capitalist, who may pressure researchers to develop products too early. This could result in a major bottleneck for R&D, one that is completely avoidable.

It's also plausible that in the near future, just a few major players will own the vast majority of CRISPR patents. In the United States, the European Union, and China, the biggest technology companies—Google, Amazon, Apple, Alibaba—have been embroiled in multiple anti-monopoly investigations and lawsuits. Do we really want to go through this again in ten years with companies that hold the keys to life-saving therapeutics and solutions to our coming global food crisis?

As of April 2021, there were more than five thousand general patents for CRISPR and more than one thousand for CRISPR-Cas9 in the United States alone. There were thirty-one thousand CRISPR patents and applications listed in the World Intellectual Property Organization's database, which aggregates IP filings from national and regional patent offices. Hundreds of new CRISPR patents are being filed every month. Now here's the rub: CRISPR is certainly the most famous technology within synthetic biology, but it's hardly the only one. CRISPR represents just a tiny fraction of R&D activity within the broader synthetic biology ecosystem.

The United States is a (mostly) free market economy where privately owned businesses and government both play roles, but where profit drives decisions. Neither researchers nor investors want to stifle innovation, but in the case of intellectual property, it's the process—not the final product—that can be owned. When the process is biological, and the end product is an organism, new genomes could become new economies.

We've already seen this play out: in May 2021, the Biden administration called for Moderna, Pfizer, and BioNTech to waive patent rights to their vaccine technology in order to open up the near-term supply of vaccines to the rest of the world. "The administration believes strongly in intellectual property protections but in service of ending this pandemic supports the waiver of these protections for COVID-19 vaccines," the announcement read. "We will actively participate in text-based negotiations with the World Trade Organization (WTO) needed to make that happen. These negotiations will take time given the consensus-based nature of the institution and the complexity of the issues involved."[36]

The real problem isn't the existence of patents and IP laws—it's that our current laws date back to America's founding in the late 1700s. They haven't been adapted to the realities of our current biotechnologies. Think of each gene—each sequence—as a new and scalable platform for productivity. In the previous chapter, we explained how Harvard researchers figured out how to engineer bacteria to store excess CO2 and nitrogen, which can be used to safely and organically fertilize crops. If this were a mechanical system—a metal device outfitted with solar panels—there would be no question about intellectual property rights. But when it comes to biological processes, the IP laws are less clear. In the bioinformation age, genetic data have intrinsic value, and the processes and organisms that will emerge over the next decade will challenge patent and trademark offices in ways they aren't prepared for.

Technically, forecasting the long-range effects of CRISPR, or any biotechnology, is outside the remit of the US Patent and Trademark Office and just about every other patent and trademark office around the world. They aren't futurists—they're mostly lawyers.

## RISK #6: THE NEXT DIGITAL DIVIDE WILL BE GENETIC

Parents, it goes without saying, want the best for their children. But consider the lengths some will go to to get their kids into better, more prestigious colleges. The former CEO of a major financial services company paid hundreds of thousands of dollars in bribes to secure admissions to elite schools for his children and was sentenced to nine months in prison.[37] A former cochairman of a major New York law firm pleaded guilty to paying $75,000 to hire someone who would take his daughter's ACT exam, and he was sentenced to a month in prison.[38] These are bright, successful people who broke the rules and cheated the system to do what they thought was best for their children.

If parents are willing to pay vast sums of money or cheat on exams to get their children into elite schools, think about what they might do to help them avoid chronic diseases or low intelligence. Wouldn't they intervene? If parents could increase their chances of giving birth to a healthy baby—or a healthy baby likely to score in the highest percentiles of intelligence or athletic ability—who among them wouldn't select for enhanced capabilities?

It is possible today to perform a complete analysis of all the fertilized eggs developed for in vitro fertilization before implantation. Typically this service, which can cost between $6,000 and $12,000 per round, is not covered by insurance. Private companies develop genetic report cards on frozen embryos, encouraging parents to choose their favorite. Genomic Prediction is one such provider. It offers a polygenic score, measuring DNA at several hundred thousand positions in order to predict the likelihood of a future person having, say, low intelligence, or being among the shortest 2 percent of the population. It uses genetic profiles of NFL quarterbacks to determine how closely the embryo matches those profiles for athletic ability. Reducing genetic uncertainty through report cards in this fashion strongly incentivizes people to choose to undergo IVF rather than opting for a natural, spontaneous pregnancy—if, that is, they can afford it.

As synthetic biology matures and IVF costs come down, the market will pressure insurers to pay for IVF. After all, it's more cost effective to select healthy embryos than to pay for a lifetime of care because of a preventable mutation. Those with excellent insurance plans, or those able and willing to pay out of pocket, could create dozens—and eventually hundreds—of embryos, and select the one likeliest to have the right combination of genetic advantages. When those babies are born, their genomes will be sequenced, and some of their umbilical cord blood (a rich source of stem cells) will be extracted for storage. As they age, these children will have access to this rich genetic trust fund that will pay continuous dividends, in the form of health information and a ready supply of genetic material.

Where does genome sequencing and genetic treatment end, and genetic enhancement begin? We now expect that, within the decade, CRISPR and other genetic tools will be developed to manage viruses, repair tissue, combat mutations, and lengthen our lifespans. China's BGI Group, one of the world's largest sequencing companies, already says that it can boost the IQ of children by up to twenty points, thanks to genetic selection. That's the difference between a kid who struggles with algebra and a kid who aces advanced calculus on her college placement exams. Intelligence is, of course, a polygenic trait. We know remarkably little about how the human brain works, and even less about the biological features and real-world experiences that translate into top-tier cognitive abilities. BGI Group is actually making a measured bet: as it sequences more people, the data will reveal patterns among those who are smartest. Then, it's just a matter of identifying genetic markers and selecting for them before implantation—or even upgrading embryos with parents' desired features.[39]

Not everyone will have access to such technology. Uninsured or underinsured people, who become pregnant in a bedroom instead of a lab, will not select their favorite embryos or have the option to upgrade. Their children will be statistically disadvantaged compared to their selected, edited, and enhanced peers. The genetic divide will become more obvious as children age, positioning tech-enhanced kids as superior to their "naturally" conceived schoolmates. Such "natural" children won't have a genetic trust fund to unlock, so, as they age, each new illness will be a mystery, and a diagnostic challenge for doctors. Of course, that's exactly how people live today. But we're living in an era when genetic knowledge isn't yet commonplace.

Technology-assisted pregnancy will also pit the world's richest and poorest countries against each other. While that's plenty fraught to begin with, let's add a key complicating factor: in wealthy nations such as Estonia, Sweden, Norway, and Denmark, where organized religion plays a diminished role in society, genetic selection and sequencing at birth will likely meet with less resistance from citizens. Technology-assisted pregnancy could become more easily accepted in such places. People in poorer countries, such as Malawi, Indonesia, and Bangladesh, will need to rely on sex for procreation. In technologically advanced, wealthy countries where religion plays a key role—such as the United Kingdom, the United States, Australia, the United Arab Emirates, Qatar, and Saudi Arabia—politicians and citizens will need to reconcile religious doctrine with the benefits of genetic selection and enhancement. Failure to take action could hamper their workforces, growth, and economic competitiveness.

## RISK #7: SYNTHETIC BIOLOGY WILL LEAD TO NEW GEOPOLITICAL CONFLICTS

Over the past decade, China has quietly created a scaled, national DNA drive to collect, sequence, and store its citizens' genetic data. DNA repositories are part of a wider panopticon, aided by the Chinese Communist Party's ambitions for artificial intelligence, to allow the government to continually surveil its constituents. In Xinjiang, in northwestern China, the program was billed as "Physicals for All," and nearly 36 million people took part in it, according to China's official news agency, Xinhua. Much of the government's early DNA initiatives centered on the Uighur population, whose data was reportedly being collected to help distinguish among China's many ethnic groups.[40] In a 2014 paper, researchers cited different genetic markers for Uighurs and Indians who live in China's far west regions bordering Kazakhstan, Kyrgyzstan, Afghanistan, Pakistan, and India. Chinese government researchers contributed the data of 2,143 Uighurs to the Allele Frequency Database, an online search platform partly funded by the US Department of Justice until 2018. The database, known as Alfred, contains DNA data from more than 700 populations around the world. This sharing of data could violate scientific norms of informed consent, because it is not clear whether the Uighurs volunteered their DNA samples to the Chinese authorities, and it's very unlikely that everyone involved knew that their DNA was being collected or understood the implications. Human rights activists say a comprehensive DNA database could be used to chase down any Uighurs who resist conforming to the campaign. Chinese officials have cited tracking down lawbreakers and criminals as a key benefit of the genetic studies. Another way to look at it is that this is a convenient way to build a gigantic genetic database.[41]

China continues to collect wider swaths of genetic data from Uighurs and other minority ethnicities, as well as from Han Chinese (who make up 91 percent of the country's population).[42] Soon it will have a comprehensive, powerful genetic dataset unrivaled by any other country's. The United States, Canada, the European Union, and the United Kingdom are debating the merits of genetic privacy. China, which collects massive amounts of information, and whose citizens seem untroubled by government surveillance, will face far less resistance to genetic studies and experimentation.

Will China decide to edit or enhance its population? It already has, crudely. In 1979, the CCP instituted a one-child policy to make sure that population growth, which was then skyrocketing, did not exceed economic development. Intended as a temporary measure, it prevented an estimated 400 million births. It also led to female infanticide; in China, there are between 30 million and 60 million "missing girls."[43] The policy officially ended in 2015, and as of July 2021 all restrictions were lifted. China may have solved a possible economic crisis, but it caused a present-day loneliness epidemic. Tens of millions of straight men can't marry. There aren't enough women.

This next time around, genetic tools will allow people to select a multitude of traits beyond gender. More than likely, this would begin as a nationwide program to screen and sequence would-be parents in advance of potential pregnancies. (BGI owns more sequencers than any other company or institution in the world.[44]) Initially, the program would be designed to identify and mitigate genetic problems, such as coronary artery disease. As government-funded screening gained wider acceptance and desirability, would engineering follow? Remember that China is a largely nonreligious country; unlike elsewhere, faith-related protests would not be a factor in advancing technology-assisted pregnancy.

In time, genetic enhancement will become acceptable. BGI might offer genetic screening for intelligence and a host of other desirable traits. This screening, combined with IVF treatments, would mean that future generations of Chinese people could be healthier and smarter than the rest of the world's citizens; they could have increased endurance, greater sensory abilities, and be far more resilient to disease. If China gained an insurmountable competitive advantage in this way, and if that information became public, would the United States do nothing to counter it?

Consider the potential repercussions. Universities in the United States might discriminate against Chinese students, fearing they would outperform their American counterparts. Or, perhaps they would recruit Chinese students, whose elite capabilities would make colleges more competitive. The US military would assess field readiness and determine that China had gained significant capabilities in hacking, psychological ops, and novel weapons research, and that the United States needed to regain its ground, and fast. This could mean asking, or compelling, American soldiers to opt in to genetic enhancement programs. Undoubtedly, all of this would be met with fierce resistance and social unrest. Faced with a known unknown—some people had been enhanced, but who and how much?—top government officials would be forced to make difficult decisions under duress. Would the United States then try to advance our population using genetic enhancement? Would the ultimate act of patriotism become having children using IVF, and trusting a genetic report card to select the best ones?

This development in turn would force a new cyber-biological arms race. It would be different from nuclear arms proliferation, which is observable: we can see countries building reactors, and we can trace raw materials being moved around the world. If bioescalation becomes the new normal, we may not know that countries are intentionally enhancing their populations until many years later, if at all.

Nonstate actors could also pose a future threat. What if a rogue community of people found doctors and scientists—someone like He Jiankui—willing to experiment on their embryos? What if those experiments were performed in a floating ocean city, outside the territory claimed by any government? What if ultra-wealthy, mega-powerful people, who would do anything to provide a bright future for their children, became part-time sea-steaders in newly created island nation-states, willing to circumvent the laws, take great risks, and make genetic enhancements?[45]

## RISK #8: SUPER-MICE AND MONKEY-HUMAN HYBRIDS

Japanese researchers at the University of Tokyo in 2017 injected mouse induced pluripotent stem cells into the embryo of a rat that had been edited to grow without a pancreas. As the rat matured, it formed a pancreas made entirely of mouse cells. The team then transplanted that pancreas back into a mouse that had been engineered to have diabetes. Remarkably, the rat surrogate had produced a fully functioning pancreas for the mouse, who, cured of diabetes, went on to live a healthy life.[46] In a more worrisome milestone in biology, in 2021 scientists at the Salk Institute for Biological Studies in La Jolla, California, grew macaque monkey embryos that were injected with human stem cells. They were allowed to grow for twenty days before being destroyed. This wasn't a rat and a mouse. This experiment involved two closely related primates.[47]

There is a term for these hybrid life forms: chimeras, named after the fire-breathing monster of Greek mythology that was part lion, part goat, and part serpent. The scientists conducting these studies hope that part-human chimeras, like those being developed at the Salk Institute, could be used someday to study various medical conditions, or to grow organs needed for transplantation. First, though, chimeras will have to be designed and genetically constructed in a lab. Synthetic biology takes us one step closer to that eventuality.

The notion of a monkey-human hybrid leaves one with a lot to absorb. It's ethically complex. To name just one reason why: at some point, chimeras will inherit qualities that are somewhere between humans, on which experimentation isn't allowed, and animals, which are often bred specifically for research. We don't have a system in place to define "human" characteristics in a world in which animal-human chimera live. How will we decide when an animal becomes *too* human? What if such chimeras escape? And cross-breed in the wild? What if they use primate strength and human-level wits against those who might be keeping them in a lab?

What if a super-predator was intentionally created by a bad actor—like a hyper-intelligent, aggressive, quadruple-muscled dog?[48]

Why would we even make chimeras? Think back to how Frederick Banting and Charles Best, whom you met in the first chapter, removed the pancreases of dogs and tried to treat them with synthesized insulin. This time around, animals could be genetically modified to grow without certain organs—such as kidneys—and then human stem cells could be edited into them so that they would grow human kidneys instead. (Which, ironically, means we could be back to growing animals as surrogates, and harvesting mass quantities of organs as needed.)

Another key way that chimeras will be used is to study biological development. Human-monkey chimeras will be developed to research the brain, in order to better understand Parkinson's and Alzheimer's. But what happens when some human-nonhuman chimeras develop mental capacities somewhere between ordinary animals and humans—if, say, a human-pig chimera develops enough to have an IQ of 39, which could qualify it as a severely mentally disabled human? We wouldn't agree to slaughter a human with a low IQ. They would be conferred rights, just as anyone else would. Could a chimera that has some level of human intelligence be used for research or organ harvesting? We don't have a way to designate a moral status for chimeras, and the rights and obligations to which they would be entitled.

Inevitably, chimera research would cross over to life enhancement, and supersede work around life preservation. Hummingbirds can see colors humans can't even imagine, including ultraviolet variations.[49] Future researchers could borrow what we know of the hummingbird genome, use AI systems to determine genetic constructs, and synthesize chimera genomes in the lab. This would be a tiny tweak, which could be performed with precision and scalability. Then, humans could see like hummingbirds. Further into the future, other chimera discoveries would follow, giving humans capacities found elsewhere in the animal kingdom, such as a bat's sonar, or the African elephant's superlative sense of smell.

Humans with chimera elements would likely need reclassification, and society would also categorize them differently. We are already struggling with issues of equality between races, ethnicities, and gender in the United States. Societies are not prepared for the psychological, moral, and ethical challenges of chimera research—and the outcomes that could result.

## RISK #9: SOCIETY WILL BREAK DOWN FROM MISINFORMATION

Science is collaborative. But we are living through a cultural moment that is intensely divided. Nationalism is on the rise. America is grappling with a racial injustice reckoning. COVID-19 led to distrust in government, science, and media. As concerning as dual use, DNA hacking, murky regulations, and bioescalation are, there is one risk looming even larger over the future of synthetic biology: misinformation.

The scale of misinformation, which is false or inaccurate information deliberately intended to deceive people, extends far beyond any one community or country. In late 2020, Facebook said that it removed 1.3 billion fake accounts. It removed more than 100 networks designed to spread fake information between 2018 and 2021.[50] The company says that it has 35,000 people working to combat the spread of misinformation.[51] That means it has more people dedicated to weeding out misinformation than the total workforces of Fortune 500 companies Fannie Mae (7,500), Conagra Brands (18,000), and Land O'Lakes (8,000) combined.[52] And Facebook is just *one* source of information.

Even before COVID-19, much of that fake information concerned science. In 2019, a viral story making the rounds online incorrectly claimed that instant noodles are linked to cancer and stroke, and as of May 26, 2021, the story was still active on Facebook.[53] That same day, a Google search for "ginger 10,000x more effective at killing cancer than chemo" (again, completely false), returned six pages of news articles, websites, and social media posts, some of which debunked the claim along with many that perpetuated it. Sometimes, the misinformation itself is a profitable business. Ty and Charlene Bollinger created a multichannel media operation to distribute disinformation about cancer, vaccines, and COVID-19.[54] For a while, they hawked packages ranging from $199 to $499 that included hundreds of hours of terrifying videos, along with booklets and news articles, intended to stoke fear and distrust via conspiracy theories. The couple has said that they sold tens of millions of dollars' worth of misinformation products.[55]

Misinformation campaigns are now causing parts of society to collapse. In the wake of a deadly global pandemic, Americans had a miraculous vaccine—yet four out of ten people refused to get it. On December

14, 2020, the first person in America—a nurse named Sandra Lindsay—received a jab in her arm to deliver the vaccine.[56] By the end of May 2021, when vaccines were free, widely available, and easy for anyone over the age of twelve to get, only a third of Americans—roughly 129 million people—had been fully vaccinated.[57] During that same period, from December to May, more than 250,000 people died from COVID-19.[58] Our trust in medicine and public health is clearly broken.

Disinformation campaigns have been equally damaging in politics. Conspiracy theories about the 2020 US presidential election ultimately resulted in thousands of people storming the US Capitol Building on January 6, 2021—the first coordinated attack on the legislature since the War of 1812. Many were injured and five people died.[59] Trust in our democratic process, and in the peaceful transition of power between administrations, had crumbled. Public trust in government is at an all-time low: 75 percent of Americans say they do not trust institutions to act in their best interests.[60]

The public's trust in science, in regulators, and in our institutions is fundamental to the social compact underpinning society. Which means that the biggest risk posed by synthetic biology is also a risk to both society and the field itself. Misinformation erodes trust and it leads to confusion about viruses, genome writing, CRISPR editing, and other biotechnologies, which could become critical to our long-term survival. We're about to share with you a story that illustrates these risks and the great promise, and ultimate peril, of a genetic editing project that was destroyed because of falsified data, deception, and mistrust. It's the story of Golden Rice, and it explains why, without trust, synthetic biology's great future may never be built.

| EIGHT |

# THE STORY OF GOLDEN RICE

LONG BEFORE WASHING MACHINES WERE UBIQUITOUS, MONDAY WAS laundry day in New Orleans. This meant that, at the beginning of every workweek, women stood by a crank and a wringer, pulling soiled clothing and linens through to clean them. If shirts and pants were especially grimy, they boiled them in a pot of water. All this took hours of arduous manual effort. Finishing it all in time to make dinner was a nearly impossible feat. But red beans, cooked with a ham hock and bits of spicy sausage, could be left to simmer all day long. Mixing them with rice made a hearty, delicious meal that was nutritious, too—the two starches combined to form a complete protein.

So Monday—washing day, as it was known—meant red beans and rice, and this traditional meal endured long after we stopped cooking the dirt out of our clothes.

Locals still eat savory red beans and rice. You can find the dish in most bars and restaurants in New Orleans, and the one on the menu at Lil' Dizzy's Café in the heart of the historic Tremé neighborhood is among the most iconic. Its owner until early 2021 was Wayne Baquet Sr., a second-generation Creole restaurateur. His father and aunt got into the business in the 1940s. Over the years, his family has fed famous touring musicians passing through town, football stars, and at least one president, but locals flock to the lunchtime buffet. After a short closure when the coronavirus hit, it's now open once again under the ownership of the third generation: Wayne Sr. sold the business to his son, Wayne Jr., and his daughter-in-law.[1]

Anyone can make red beans and rice—it's just two staples in a pot. But to make *great* red beans and rice you need the Holy Trinity (diced onion, celery, and bell pepper) precisely balanced with sage, parsley, minced cloves, Andouille sausage, and ham hock. Even so, what really makes the dish isn't the beans. It's the rice.

In the United States, most of our rice is grown in the South. Louisiana cultivates 2.7 billion pounds of it a year, which is worth around $360 million. More than half the world's population relies on rice as a primary daily staple. But the most popular variety, white, is stripped of its whole grain, which contains fiber, minerals, vitamins, and antioxidants. So rice, for much of the world, is filling, but not very nutritious. You can blame Confucius, in part, for our aversion to healthier brown rice. Late in life, he decided that whole grains were for the unrefined masses. Rice could "never be too white," he declared; that made it the correct and proper backdrop for bright green vegetables.[2]

Rice was first domesticated nearly ten thousand years ago near China's Yangtze River. Back then, it was a good source of iron, fat, fiber, potassium, calcium, B vitamins, and manganese, a mineral that doesn't get a lot of attention, but which supports our blood sugar regulation, nerve function, and bone development. In the thousands of years that followed, the human diaspora spread. As people migrated to new territories, they brought seeds with them and introduced genetic modifications, taking advantage of natural mutations to breed new shapes and textures that grew better in different locales and climates. Just about all those early modifications resulted in a lighter, whiter rice. Polished short-grain sticky rice became popular in Japan, while in Pakistan and Jordan, long-grain basmati rice was preferred.[3]

The long-grain rice served at Lil' Dizzy's—and the rice you buy at the grocery store to cook at home—is the result of thousands of years of careful engineering by farmers. Clever marketers might use "ancient grains" on their packaging, but all grains today have been modified through *cisgenic* breeding, in which genes from the same or a closely related species are inserted to improve yield, confer greater tolerance to drought and heat, and increase nutritional value. Essentially, cisgenic breeding uses new techniques to speed up processes that could have been done with conventional breeding methods. In the United States, much of the rice we consume has been modified twice: once through careful breeding, and again when it is enriched with iron, folic acid, niacin, thiamin, and other nutrients that were lost when the rice was processed to make it white. Even after being enriched, though, this rice isn't a great source of vitamins and minerals. That's fine in New Orleans, because the local diet is already full of many foods packed with essential nutrients—including the red beans and collard greens served up at Lil' Dizzy's buffet. But that's not the case everywhere else.

Every time you walk into a restaurant, you enter into certain unspoken agreements with the owner, cooks, and support staff: the food being served is fresh, free of disease, and prepared safely. Trust is paramount. What you eat is the result of myriad decisions made in the kitchen, in the supply chain, and out in the fields, where raw ingredients like rice are grown. The locals and tourists who show up at Lil' Dizzy's know how renowned the restaurant is and how long the Baquets have been cooking. Trust is established. Still, the tourists often want to hear how spicy the rice and beans are, just in case. They also want to hear about the history of the restaurant, and maybe the family stories behind the recipes. Most probably don't ask about how the rice is grown; after they leave Lil' Dizzy's, it's unlikely that they'll rush to research the thousands of years of careful cultivation that resulted in the perfectly cooked long-grain rice on their plates. Or to go even deeper and research something that sounds like fiction but isn't: a long-running quest to transform simple, everyday rice into a global superhero, one that could save the lives of millions of malnourished people each year.

But if they did, they'd hear a harrowing tale about two plant biologists who created a new strain of rice intended to feed the world's poorest and a global misinformation campaign designed to foment distrust in science and discredit their work. They would learn that decades of peer-reviewed scientific research, meticulous testing, and observance of established protocols to reduce risks—essentially doing everything right—could be derailed by misinformation. Great scientific achievements must be accompanied by even greater efforts to gain the public's trust. That story begins in a field trial about eighty miles northwest of Lil' Dizzy's, and results in a group of violent activists tearing apart a rice paddy in the Philippines.

## THE PROBLEM WITH RICE

Rice is simple. Cultivating it is not. Amy once lived in northern Japan, where, adjacent to her house, there was a small rice paddy owned by a local family who tended it as a hobby. Every spring, she helped them flood the paddy, which resulted in about six inches of standing water covering the ground. They then took trays of rice seedlings and plunked them down, one by one, into neat rows. During the summer growing season, the water level had to be maintained—too much water rotted the grain, while too little meant the rice grasses and hulls dried out. In the fall, they'd drain the paddy and gently beat the grains, to help them dry more quickly. It was too small an area for a harvesting machine, so in early October, they did the work by hand, cutting the plants using a sickle. (This was back-breaking, hamstring-shredding work for Amy, but not much of a challenge for the family.) Bundles of rice were tightly bound and lined up on the ground in shapes that resembled brooms taken off their handles. Eventually those bundles would be hung over temporary wooden fences, to allow gravity to pull the remaining drops of water out as the grains dried in the sun.

Growing rice the traditional way, Amy came to learn, requires knowledge, skill, luck, and some serious hard work. Even under the best circumstances, the yields aren't great: a square foot of land only accommodates ten rice plants, and each one produces only seventy to one hundred grains of rice. The family whose rice paddy Amy helped tend was affluent, so if any summer was too wet or too dry to produce a good crop, it wasn't catastrophic: they'd just buy their rice at the grocery store. But millions of people who cultivate rice—small-scale farmers and poor families that grow their own—can't do that.

Beginning in the 1960s, distinguished researcher and plant biologist Ingo Potrykus dreamed of solving two problems with rice: improving its nutrition and making it easier and more predictable to grow. As a scientist, Potrykus knew how critical key nutrients were to childhood development. He also knew what it was like to go to bed hungry. He had lost his father during World War II; in the war's aftermath, Potrykus and his family had fled East Germany, and he and his brothers had pilfered shops and begged on the street, trying to scrounge up any food they could.[4]

Potrykus could engineer rice to double or even triple yields, but that wouldn't solve the nutritional problem. He knew that rice was readily available in many communities, but while people who lived in them seemed to have enough food, they were severely malnourished. So Potrykus began ruminating on the structure of the plant and the feasibility of adding to its genetic code. One night he drifted off to sleep while thinking of different possibilities. He woke up having pieced together some new hypotheses: Could rice be enriched with fiber or potassium? What about crossing spinach genes with rice?

Potrykus needed a plant source that when spliced together with rice would enrich the grain with critical nutrients without changing its flavor, texture, or density—and, most critically, wouldn't require different growing techniques. He landed on plants that are high in carotenoids, which help them absorb the light energy they need for photosynthesis. Plants with especially high levels of carotenoids are vivid shades of red, yellow, and orange. There are more than six hundred known types of carotenoids, but the one you're most familiar with is beta-carotene, which is abundant in carrots, squash, sweet potatoes, mangos, grapefruit, bell peppers, and tomatoes. Beta-carotene also acts as an antioxidant, has strong cancer-fighting properties, and is metabolized into vitamin A.

That last part is key. Vitamin A deficiency afflicts millions, and it wreaks all sorts of havoc on their bodies and health. The old adage about eating carrots to help your eyesight is partially true. Extra vitamin A won't correct your nearsightedness, but a vitamin A deficiency will cause disastrous optical, neurological, and immunological problems. Too little vitamin A, and the cornea, which has the consistency of a sponge, can't dry itself out. We have layers of cells on our cornea that are continuously pumping it dry, and they need vitamin A to perform. If they can't, a milky white sheen will cover the iris. Without treatment, it turns opaque—

making the eye look completely white and rendering vision blurry. Of all the consequences to vitamin A deficiency, whited-out eyes and blurred vision are probably your best outcomes.

For instance, if you don't get enough vitamin A, over time your cornea begins to erode. There aren't enough healthy, preprogrammed cells ready to battle an injury. Without the cornea protecting the front surface of your eye, the nerves behind it are exposed. Even without, say, getting hit in the eye with a ball, corneal erosion feels like being struck in the eye with a hot poker, over and over again. Optometrists describe the sensation as "religion-altering pain," because, as one put it, it will make you "pray to any god to make it stop."[5] If you were to examine the eyes of someone suffering from long-term vitamin A depletion, you'd also be likely to find irreversible signs of blindness.

Potrykus knew that widespread vitamin A deficiency meant that excruciating eye pain, and possibly a complete loss of sight, was inevitable for hundreds of millions of people. He also knew that too little vitamin A led to compromised immune systems, and that, in such cases, children often suffer the most. Even a mild vitamin A deficiency dramatically increases childhood mortality rates, because it reduces resistance to infectious diseases such as measles as well as ones that cause diarrhea. In some impoverished communities, the childhood mortality resulting from too little vitamin A can exceed 50 percent.

Potrykus considered the options. The public health sector might develop a concentrated vitamin A serum at just the right dosage to be effective over many years—potentially tricky, as too much would be toxic—and then convince billions of people around the world to line up for an injection. Or, even though this presented an enormous scientific challenge, he could give rice a beta-carotene boost.

The rice plant species most commonly consumed is *Oryza sativa*, which only has 12 chromosomes and a total of 430 megabases, which is a nucleotide length of 1 million base pairs. This makes it an excellent candidate for plant genomics. On its own, *Oryza sativa* does not express beta-carotene in the starchy, interior part of the rice grain (known as the endosperm) that most people eat. Potrykus started with the hypothesis that a new beta-carotene pathway could be engineered into the plant.

Potrykus and a small team of colleagues started working on the idea. He wasn't the only scientist thinking about beta-carotene and rice. The Rockefeller Foundation, a private foundation based in New York City with more than $4 billion in its endowment, and a primary mission of ending global hunger, had been mulling the same problem, and had also landed on the idea for vitamin A–enriched rice. It became an early investor in the International Rice Research Institute (IRRI), a nonprofit scientific center headquartered in the Philippines. By 1984, Rockefeller's director of food security, Gary Toenniessen, and some of his fellow program officers felt they had all they needed for a sweeping global program aimed at developing a new super-rice: internal expertise, a network of institutions and partners, and the means to recruit outside scientists. Rockefeller's scientists soon created the first DNA molecular marker map of rice, and eventually they connected rice to the evolution of cereal grains such as maize, rye, and wheat—a stunning finding that challenged cherished beliefs about other key sources of food. But enriching rice to make it more nutritious? That still eluded them.[6]

Toenniessen decided to host a meeting with a biochemical researcher in the network, Peter Beyer, an expert on beta-carotene at the University of Freiburg in Germany. While Toenniessen and Potrykus thought it would be possible to engineer a new strain of rice that would be nutritious and easy to cultivate, they still needed outside genetic code. Bell peppers and sweet potatoes were good sources of beta-carotene, but Beyer hypothesized that a more distant plant relative would be a better candidate. One that, while beautiful, was rarely mistaken by humans for food was the daffodil, the familiar bright yellow flower with six petals that surround an orangey-yellow trumpet.[7]

It was a ludicrous idea, they knew. In order to genetically modify *Oryza sativa*, they would first have to determine which daffodil genes to use. Next, they would need to isolate the genes and encode them into plant embryos. Assuming they selected the correct genes, theoretically the embryonic plants would incorporate those new genes into their DNA as they grew, producing the desired proteins and operating in concert with the rest of the plant's genome. Eventually, those cells should mature and produce seeds that carried the newly modified code; the plants that grew from those seeds would then pass the new genes down to future rice plants capable of producing beta-carotene.

That was just one piece of the puzzle. Historically, plant breeders have relied on brute force experiments and patience. By the 1990s, scientists were able to transfer a single gene to modify an organism. But beta-carotene-enriched rice needed three modified genes. Potrykus and his team began trying different methods for the transformation. At first, they planned to introduce one new gene each into individual rice plants, and then breed them conventionally. While a few early test plants looked promising, repeated attempts to produce all of the needed enzymes failed. Then, they tried a more aggressive approach: they would engineer a bacteria that could insert its DNA into the targeted rice plant embryos. This process, known as an *Agrobacterium*-mediated transformation, would introduce all the genes necessary at once. The newly modified DNA would include phytoene synthase and lycopene beta-cyclase from the daffodil and phytoene desaturase from the bacteria. When fully mature, these engineered rice plants would produce and store beta-carotene.

Rice was just as challenging to grow inside Potrykus's greenhouse on the foothills of the Swiss Alps as it is in a Japanese rice paddy. It would take many years of trial and error before Potrykus, Beyer, and their teams would publish their research describing a bioengineered rice enriched with a vitamin A biosynthetic pathway. But academic papers were only the beginning. This bioengineered rice needed to be studied and tested in rice paddies around the world, and that would take many more years of dogged research. The team's ultimate goal wasn't commercialization, but rather a global seed giveaway: a humanitarian project that would someday distribute their bioengineered rice for free to every farmer and every family who needed it. New genetically modified crops of enhanced rice would mean that no child would die from a vitamin A deficiency ever again.[8]

Around this time, the public was first learning about genetically modified organisms. The Flavr Savr tomato had inspired other researchers to address extant problems—ring spots on papaya, apples that didn't bruise as easily—and their work attracted the attention of activists from around the world. One group in particular, Greenpeace, took an active role in discrediting the science behind genetically engineering plants. They were willing to get arrested, if it contributed to a ban on GMOs—especially if it resulted in media coverage.

By now, Potrykus and Beyer had been laser-focused on their project and the prospect of its world-changing outcome for food insecurity for years. They'd been steeped in the world of pure science for even longer. But unlike the Baquets at Lil' Dizzy's, they didn't know that winning the public's trust requires a certain choreography, and they didn't have long-time ties rooting them in the communities they sought to serve, or a heart-warming family story to tell. They, and those who supported and funded them, couldn't foresee that millions would soon lunge for reasons—any reasons, be they real or wholly concocted—to oppose them and the science behind synthetic biology.

## THE GIANTS AND THE GOLDEN BACKLASH

News about this new rice first came to the public's attention with the publication in January 2000 of an article in *Science* detailing the team's latest research breakthroughs.[9] By now, public debate about GMOs and genetic engineering had been escalating for a decade, so the journal decided to send copies to 1,700 journalists worldwide with a special editorial note: "This application of plant genetic engineering to ameliorate human misery without regard to short-term profit will restore this technology to political acceptability." It was a remarkable effort to get ahead of misinformation and salacious headlines.

For a moment, that effort worked: public discussion about this latest, heroic bit of genetic engineering, and the implications of the futuristic technology, landed well. The strain became known as Golden Rice, both for its mango-like hue and its potential value to human society.

Biotechnology research is vital to human progress, but it's expensive. Outside investors or large corporate concerns often foot the bill, and it might be decades before they ever see a return—if one comes at all. Patents are one way to balance the extraordinary risk in biotech R&D. Not everyone follows intellectual property rules, however, as we explained in the previous chapter, and legal challenges are common. Courts still haven't determined how much a gene sequence must be altered in order to qualify as a "patentable invention," or what constitutes a patent infringement when academic researchers use patented genomic material.

By April 2000, as Potrykus and Beyer were readying their bioengineered rice for its first field trials, the IRRI had asked for a patent search. The legal review determined that Golden Rice had been produced using somewhere between 70 and 105 patents, licenses, and other binding legal agreements—which were not controlled by one entity, but instead held by more than 30 public and private institutions. An additional complication exists: each country recognizes patents differently, which raised the plausible threat of years of unwieldy legal battles if Golden Rice was produced and distributed. Golden Rice might have been a scientific success, but from an intellectual property standpoint, it was an utter disaster.

Understandably, the team was devastated. Nearly two decades of research had produced a viable biotechnology, but it might never be allowed to leave the lab. That's why, when some pharmaceutical representatives contacted Potrykus and Beyer, they took the meeting.

The enormous pharmaceutical corporation AstraZeneca held some of the patents that Potrykus and the team had used to create the Golden Rice. AstraZeneca offered them a deal: the company would settle the team's existing IP challenges, license the patents and technology back to them free of charge, and continue to fund the team's work. The seeds would be distributed for free to farmers earning less than US$10,000 a year. But there was a catch. A division of AstraZeneca, Zeneca Agrochemicals, and the agricultural business of the pharmaceutical giant Novartis were merging to form the biotechnology company Syngenta—it would soon become the world's largest global provider of seeds and agrochemicals, as well as a major player in genomic research. In exchange for enabling the team to continue its work, Syngenta would gain the marketing rights to Golden Rice and the right to sell the seeds commercially.

Potrykus and Beyer understood the optics. What *Science* had praised the team for—developing the rice biotechnology without regard to short-term profit—would surely be placed in doubt by the deal. They'd received $100 million in public and private funding with the promise to feed the world's most malnourished, and now they were taking their research, IP, and expertise to a giant corporation that would surely profit off the poor.

But they felt they had to accept AstraZeneca's deal in order to see the project to completion.

Criticism came swiftly, and it was harsh. Rural Advancement Foundation International, an advocacy group based in Winnipeg, Canada, called it "a rip-off of the public trust," adding, "Asian farmers get (unproved) genetically modified rice, and AstraZeneca gets the 'gold.'"[10]

On the advice of AstraZeneca, Potrykus and Beyer agreed to a media campaign of news conferences and interviews with leading publications. At one news conference in New York City that May, Beyer joined Robert Woods, who was then AstraZeneca's president, to announce that Golden Rice would be made available worldwide within three years. It hadn't even been piloted in real-world conditions at scale, and critics quickly pointed out that the rice hadn't been tested properly, but Woods dismissed their concerns. "If we do the right things in verifying the safety," he said, "it will take out the political and emotional issues that have surrounded biotechnology."[11]

A Golden Rice Humanitarian Board was created in an attempt to restore public trust. The board would oversee development of the technology and manage noncommercial licenses to public research institutes. It would also build a network of scientists and research organizations to scale the cultivation of Golden Rice and adapt it to local growing conditions.

The new press strategy paid off. Soon, Golden Rice was being written and talked about everywhere: in the BBC, in an alternative weekly newspaper published in Los Angeles, on the pages of Live Journal, and elsewhere.[12,13] On July 31, 2000, *Time* magazine put Potrykus on its cover with an unmissable, all-caps headline in an enormous bold typeface: **THIS RICE COULD SAVE A MILLION KIDS A YEAR.** Days after the *Time* cover was published, Monsanto—the American agrochemical and agricultural biotech behemoth—announced that it, too, was developing a golden rice, and it would be giving away free licenses and other genetically engineered technologies for use by farmers in poor communities. It also proclaimed that Monsanto would release its own genomic sequence on its brand-new website: Rice-research.org. "We want to minimize the time and expenditure that might be associated with obtaining licenses needed to bring golden rice to farmers and the people in dire need of this vitamin in developing countries," said Monsanto's chief executive, Hendrik Vefaillie.

Of course, research was a very long way from completion—and Monsanto's sudden appearance didn't exactly mitigate the public's mistrust.[14]

Any new groundbreaking technology brings with it a tremendous amount of misplaced optimism and fear. That was certainly true of Golden Rice. All that initial media coverage intimated that Golden Rice was a finished technology—but it still required lab testing and additional tweaking, not to mention several growing seasons to test and perfect its cultivation in the wild. Enormous amounts of data still had to be collected and analyzed. Even if the patent issues were addressed, there were still massive regulatory hurdles to jump, such as the new Cartagena Protocol on Biosafety to the Convention on Biological Diversity, which we described in the previous chapter. The protocol only gave countries the right to bar imports of live genetically modified organisms—they could choose not to exercise that right, or ask an exporting country to provide a risk assessment of the organism, which would be performed not by an independent third party, but by the exporting country itself.

No one knew yet how to manufacture viable Golden Rice seeds at a global scale, or how best to distribute and track them. And, not least, there were historical preferences to consider: this fortified rice would be nutritious, yes—but it would no longer be white. Educating farmers and the public, gaining their trust in the crop as something that was safe to eat, and that would taste no different from the rice they'd eaten their whole lives, would be a major challenge. To put it bluntly, the science was meticulously planned—but no one had really planned a strategy for introducing it into society in a way that would garner acceptance and trust.

Few of these details were mentioned in the news, which created an opening for organizations that were adamantly opposed to any and all genetic modification. Like Greenpeace, for example, which developed a campaign to target Golden Rice. The organization used statistics to tell a very different version of Golden Rice's vitamin A story. Greenpeace issued a statement saying that the average malnourished child would have to eat fifteen or more bowls of rice every day to get enough vitamin A, and that an adult would need to eat twenty *pounds* of rice a day. There was no factual basis for these claims, and Greenpeace didn't offer any scientific explanation showing how they arrived at that number. But the science mattered less than the storytelling. A lengthy academic exposition with chemical compounds and graphs couldn't compete with a simple number anyone could immediately visualize. Twenty pounds is four sacks of standard-issue grocery store flour, or roughly twenty boxes of corn flakes—and a stupidly large amount of rice. The statement was, unfortunately, calculated to stick in people's minds—and it did, immediately fomenting doubt about the project.[15]

If very little vitamin A would be delivered through this engineered rice, some fevered minds began to theorize, then the rice must be a Trojan horse, a way to coerce and gain control of small-scale farmers, who were already being hijacked into using GMO seeds and high-priced herbicides. Free rice today would be monetized for a profit tomorrow, and it would unlock new ways for the big agricultural and pharmaceutical corporations—which, even on their best days, were never the most popular companies in the S&P 500—to make money off the backs of poor, unwitting farmers. The hype about Golden Rice was totally wrong, Greenpeace began to argue. It wasn't created to help the poor—it was engineered to sell genetically modified seeds and the proprietary herbicides they would need to grow. Word spread fast, and backlash ensued, especially in activist circles in Europe and North America, and even in small rural communities throughout Southeast Asia, the exact places that most needed Golden Rice.

Misinformation campaigns are most effective when they have some tangential connection to the truth and play on existing anxieties. That was certainly true when it came to Monsanto, a pioneer in genetically engineering crops at scale, having identified a gene that made certain crops immune to Roundup, the company's weed killer. If a farmer planted Monsanto-branded soybeans, corn, and cotton, those crops could be safely sprayed with the pesticide, which would kill the weeds but bring no harm to the plants. Of course, it also meant that small-scale seed providers got squeezed out of business, and over time Monsanto gained a lucrative position in the market.

There was another problem. The company had begun selling its seeds in Europe and the United Kingdom, which was then still reeling from an epidemic of bovine spongiform encephalopathy (BSE), or mad cow disease. Cows were falling ill with a debilitating neurodegenerative disease that spread easily on industrial farms. The British government initially told people that the highly infectious and extremely dangerous disease posed no risk to humans, and that they could continue eating infected meat. But then hundreds of people contracted the human form of the disease by eating the nerve tissue of infected cattle. Millions of people lost their trust in government to regulate industrial agriculture, and British consumers began rebelling against GMOs, and against Monsanto in particular, incorrectly conflating BSE with genetic engineering.[16]

Disturbing footage of cows shaking uncontrollably, unable to carry their own weight, unable to walk more than a few steps as they slowly died, had already run on TVs worldwide, and now there were similarly heart-wrenching videos of people, who were once strong and vibrant, suddenly bedridden and shaking, with their mouths agape and eyes vacant. For decades, people had been consuming meat without really questioning where it came from. Now, for the first time, they learned that young calves were fed a bovine meat-and-bone mixture. The disease was spreading among the herds because, as older cows became infected with BSE, they were being slaughtered and rendered into food for those who were still living.

Bovine spongiform encephalopathy is caused by proteins called prions that, for reasons we still do not fully understand, morph from healthy to harmful. Scientists believe that this happens spontaneously, like many other biological occurrences. But regardless of the cause, it was the scary stories, not the science, that grabbed public attention. Now, a government that had lied about a life-threatening infection, and corporations that were intentionally modifying seeds to reap higher profits, wanted to bring genetically modified rice into poor communities, activists asked? Did this mean that AstraZeneca and Monsanto were performing experiments on needy children? What if the engineered proteins in Golden Rice mutated? What if the rice grew uncontrollably, choking off the healthy plants people needed for survival? Questions begat conspiracy theories. People talked about future GMOs being developed in clandestine labs, and secret cabals of scientists and executives working together to grab control of the global food supply.

The activists spreading misinformation about Golden Rice were highly educated, well-read, worldly people. But they willfully ignored the science on the rice and distorted evidence in order to support their agendas. It's hard, and complicated, to explain that modern R&D is, unfortunately, inextricably tied to onerous international patent and trademark systems. This let Golden Rice's opponents peddle fear rather than facts, and led the public to arrive at snap answers to questions that the scientists were still investigating.

## GOLDEN RICE TODAY

By 2013, Golden Rice was finally undergoing public-sector trials in a few rice paddies, supervised by the IRRI and a few other partners. On one bright, humid morning in August, researchers gathered at a test site in the Bicol region of the Philippines, around two hundred miles southeast of Manila, hoping to see tiny, yellow-tinged grains of rice nestled among the tops of tall grasses. After years of legal and regulatory hurdles, and a ferocious battle against the backlash that Greenpeace had gleefully stoked, Potrykus and Beyer would at last see a viable crop of Golden Rice in the wild. It would mark the end of their many years championing the promise of Golden Rice—and the beginning of a new era in which bioengineering could help fight malnutrition on a global scale.

But on the far edge of the field was a small group of protesters, who identified themselves as local farmers. (They weren't.) They pushed over a flimsy bamboo fence and forced their way into the paddy, trampling and uprooting the plants until the entire field was decimated. The local agriculture department later reported that the surprise attack was staged by an extremist group. It had formed around a conspiracy theory that Golden Rice had been created to facilitate a multinational takeover of the Filipino rice market.[17]

George Church later decried the day's events, saying, "A million lives are at stake every year due to vitamin A deficiency, and Golden Rice was basically ready for use in 2002. Every year that you delay it, that's another million people dead. That's mass murder on a high scale."[18]

More than one hundred Nobel laureates signed a letter demanding that Greenpeace end its opposition to genetically modified organisms, stating, "We urge Greenpeace and its supporters to re-examine the experience of farmers and consumers worldwide with crops and foods improved through biotechnology, recognize the findings of authoritative scientific bodies and regulatory agencies, and abandon their campaign against 'GMOs' in general and Golden Rice in particular."[19]

But Golden Rice remained in limbo. It wasn't until December 2019 that the Philippine government issued a biosafety permit allowing the IRRI to restart Golden Rice trials, thus clearing a narrow path for engineered rice to be used as food or feed. But it will still need to win approval for commercial production before it can be made available to the public. Regulatory approvals are still pending in New Zealand, Canada, and the United States. Where the rice is most needed, though, there's been very little movement.

If we're not careful, the Golden Rice debacle will someday look like a quaint, provincial scuffle among bureaucrats. Consider, now, all you've read about the future of synthetic biology. IVG will be used to create sperm and egg cells in a lab from any adult cell, and artificial intelligence will help select the best possible embryo to implant—possibly in an artificial womb housed in a medical center, rather than in a person. Researchers will de-extinct woolly mammoths by mixing their genomes with that of the Asian elephant, and similar experiments will bring back other species from the dead. We will grow thick, juicy steaks from a slurry of stem cells cultured inside a bioreactor—and we might cross that tissue with another plant or animal to enhance its flavor and texture. There is tremendous opportunity to improve on evolution and life as we know it, but without deeply investing in education and working extraordinarily hard to curb the spread of misinformation, scientists may never earn the public's trust in their research.

## WHY WE TRUST SCIENTISTS, BUT NOT SCIENCE ITSELF

Pew Research asked members of the American Association for the Advancement of Science (AAAS) and a wide swath of the public a series of questions in 2020. The questions were about values and trust. They ranged from inquiries about what people believed about childhood vaccines to how they felt about biotechnology, from their opinions on animal research to their view of the International Space Station (ISS). The public respondents and the scientists agreed on many topics, including this ISS: 68 percent of the scientists and 64 percent of the public said that they thought it was a valuable investment. But they disagreed to an alarming degree on genetically modified foods. While 88 percent of scientists said they thought genetically modified foods were perfectly safe to cultivate and eat, only 37 percent of the public thought the same.[20]

That same study asked the public which professions it deemed most trustworthy in the United States. Scientists came in second (the military was first). Religious leaders, kindergarten teachers, and many others came in lower in the rankings.[21]

Why do we trust scientists, but not the science itself?

One reason is that we're biologically wired against changing our minds when our cherished beliefs are challenged. When you hear new information, you process it based on what you already think. Conforming new information into what someone already believes requires far less brainpower than adopting a wholly new belief, and—more importantly—people don't want to feel the shame and embarrassment of admitting they were wrong. In such instances, people instinctively use logical reasoning and critical thinking to produce irrefutable counterarguments. Studies have shown that the more educated someone is, the better they are at convincing themselves that evidence opposing their beliefs is wrong. Which means that if you began reading this chapter with a deeply held opinion on genetically modified foods—or if you already knew one side of the Golden Rice story—you're going to have to keep a very open mind as you think about the promise of synthetic biology and what follows in Part Three. That's because we will be exploring future scenarios spanning the next fifty years of planetwide evolution.

Most people find comfort in answers, and we hate ambiguity. We are emotionally addicted to certainty, which is why we convince ourselves that there is an explanation for everything. This thinking, while understandable, clouds our judgment. When faced with deep uncertainty about the outcomes of a complex topic, such as genetically engineered living organisms, feelings of anxiety and doubt can drive the narrative and make the potential outcomes seem more dystopian than they actually are. This happens even when the existing outcome—in this case, millions of people suffering or dying each year from an easily preventable malady—is more than dystopian enough.

In an alternate version of history, Golden Rice could have seen a different outcome. Patent holders could have agreed to license their use for a global humanitarian effort, and they might have agreed to charge very low fees or none at all. There might have been a campaign to build bio-literacy among the general public, making the research easy to understand and available in many different formats and languages. Part of that bio-literacy movement could have included high-profile public service announcements, with affable, trusted public figures—Michael Jordan, Oprah Winfrey, Tom Hanks—extolling the virtues of Golden Rice. Perhaps they'd have posed holding a small bowl of rice, with a spoon in hand, ready to tuck in for a hearty meal. Pantone might have made "Gold" its color of the year. There might have been a *Friends* episode, "The One Where Ross Eats Golden Rice," where the gang tries and fails to make dinner over and over again. Today, Golden Rice might be such a common staple that no one would think it was special or unusual—except for those children and their families who remembered a generation earlier suffering from nutritional deficiencies.

# PART THREE | Futures

| NINE |

# EXPLORING THE RECENTLY PLAUSIBLE

As the tools of biotechnology become more accessible, and applications for synthetic biology seep into all of our major industries, life and the pace of evolution will change. Unimagined social, economic, and security problems will arise, as will previously implausible solutions to combat hunger, disease, and climate change. An unknowable number of variables are in play: individual researchers toiling away in their labs in China, the United States, France, Germany, Israel, the United Arab Emirates, and Japan; venture capitalists and other investors analyzing the startups and determining which ones to back; companies specializing in various applications—such as Ginkgo Bioworks, which designs custom organisms—going public or getting close to going public; regulators rethinking their frameworks; and other factors. There is, of course, no way to calculate the statistical probability of the next big breakthrough, though one of Google's AI divisions, DeepMind, is hard at work improving its protein-folding algorithms, and many companies are developing a one-shot flu vaccine that would provide lifetime immunity.

Because synthetic biology is still at the Alexander Graham Bell–Chickering Hall stage of development, some might argue that there's no use in planning for the future. Any strategies would need to be revisited, and besides, there are more immediate issues to worry about, including cyberattacks, or unemployment. But it is precisely because there is so much uncertainty, and because difficult-to-reverse decisions are being made every day, that we must challenge our cherished beliefs. If we encourage "What if?" questions today, we can avoid "What now?" questions in the future. And these questions cover a broad spectrum of concerns:

- What if scientists working in these areas of research cannot develop the necessary structures to build and maintain the public's trust?
- What if the future of life is concentrated among a small pool of decision makers? Some people, because of their skill sets and knowledge, will have a greater degree of control over the evolution of life on our planet—and possibly the evolution of our species elsewhere in our solar system—than others. Who gets to rewrite life?
- What if intentional design changes our attitudes about families and parenting?
- What if, in the future, some people "own" others' genetic data? Many countries, including the United States, have ugly histories in which enslaved people were legally designated as property. Could you, or your company, own future people, by owning the rights to their genetic code?
- What if your body is hackable? What if a bad actor designs a probiotic or a virus designed to cause you debilitating gastrointestinal distress? What if a DNA registry sells your data to a third party, without your express consent? What constitutes genetic privacy? Do you have the rights to keep your genetic data private and secure from third parties?
- What if we decide that some upgrades to our basic biology should be allowed? Who should be the ones to decide what these allowable upgrades are? What if that means we wind up with new human-animal chimeras, like people who have the long, strong fingers of a macaque monkey?
- What if wealthy people pay to upgrade their babies, but most people cannot afford the procedure? How could that lack of access further

divide societies in the future? Would society discriminate against non-engineered people?

- What if politics and misinformation thwart progress in agriculture? How will we then contend with limited food supplies in the wake of our global climate emergency?
- What if countries make decisions about designing life that, while not running afoul of established weapons treaties, nonetheless do not serve the public good in the long term?
- What if China dominates both artificial intelligence and synthetic biology and creates the global standards that govern these technologies in the future? Will the United States then be locked out of key technologies and fall behind its key geopolitical rival?

Most governments have no blueprint articulating long-term research and development funding targets at this critical juncture for emerging science and technology, which includes synthetic biology as well as the adjacent technologies supporting it, such as artificial intelligence, home automation, biometric data collection, and the like. Researchers are constantly making new discoveries and developing new applications, widening the gap between the technology and our ability to assert any meaningful guidance and control over it. Governmental agencies have largely ignored this reality, and the outdated regulatory frameworks have created confusion. In the United States, science and technology policy is intertwined with politics, so as new administrations take the Oval Office, and members of Congress come and go, there is little opportunity to develop, and champion, consistent points of view, let alone norms and standards. A failure to develop strategies for emerging science and technology will result in untenable outcomes pitting our legal and governing systems against the private sector. Waiting until synthetic biology hits the mainstream all but guarantees that the United States and other developed nations will fall behind China, where it is already a core focus of national strategic interests.

US governing and regulatory systems incentivize short-termism. But the dangers of short-termism were made abundantly clear during the COVID-19 pandemic around the world, too, as governments hesitated when they might have taken decisive action. Instead of mandating safety measures, they worried about reelection and kowtowed to popular opinion. Brazil's president, Jair Bolsonaro, an early and ardent virus denier, threw out the country's previous public health policies, leaving the country with no coordinated response so that it endured wave after wave of vicious infections. By May 2021, nearly 500,000 people had died there.[1] India's prime minister, Narendra Modi, initially refused to take action on COVID-19, then abruptly shut the country down, which caused massive economic turmoil for huge swaths of the workforce.[2] Then, in another weirdly spontaneous announcement a few months later, he declared the pandemic over. India, he said, had "saved humanity from a big disaster."[3] Cricket matches resumed, parades and religious ceremonies were allowed at full capacity, and political rallies for Modi's Hindu nationalist party were encouraged. Vaccines were not available, and unlike other countries, India hadn't locked down its borders. There was no national stockpile of respirators that could be distributed in the event of a new wave of cases. Soon enough, people started getting sick. There was no plan, there was no public health guidance, and misinformation spread fast on WhatsApp and other social media platforms. Promises of false cures, misleading stories about vaccine side effects, and obviously false and racist claims that Muslims were behind the spread of the virus caused India's late-stage COVID-19 catastrophe, in which hundreds of thousands of people died in a matter of weeks.[4] But we've also seen the ruinous damage caused by short-termism in the private sector, whether that's cutting corners on safety, knowingly making products that cause addiction or harm, or prioritizing profits above what's best for society.

All of which is ample reason to face those "What if?" questions now. Doing so does not mean putting a stop to synthetic biology research—nor does it hamper innovation. On the contrary. If we engage in rational conversations now about the next-order impacts of synthetic biology, we will be better positioned to ensure that its greatest potential values—societally and financially—can be achieved. One way to do this is by creating and thinking through scenarios that play out possible decisions, actions, and outcomes. Scenarios describe how the world may develop, given what we know to be true today. If we begin with a set of "What if?" questions based on the evidence of current trends in science, along with a set of further assumptions about society, we can develop plausible responses. Say, for example, that we started with "What ifs?" about embryonic research, and put this into context with a set of assumptions about public opinion, the economy, and the like:

- What if scientists (1) created synthetic mouse embryos with pluripotent cells; and (2) focused their research on how to use in vitro gametogenesis to reverse-engineer any tissue or cell into becoming induced pluripotent stem cells?
- If so, we will assume that (1) market demand for reproductive assistance continues to rise, especially as people wait longer to start families; (2) acceptance of CRISPR to edit embryos and greater accessibility to IVF grow; (3) wealth disparity increases; (4) challenging job market conditions for Millennials and Generation Z persist; and (5) technology companies continue to nudge consumers toward quantified health trackers.

Under these conditions, what might the future look like in the next ten to fifty years?

Scenarios are an effective strategic tool used by executives and their teams to confront deep uncertainty. They are opportunities to rehearse the future. Boards of directors and executive management teams use scenarios to figure out where they should play, and where and how they can win, and to understand the assumptions that must hold true for their current strategies to succeed. Military strategists also use them, to analyze the likely consequences of different actions and strategies. Design teams use scenarios in order to anticipate new products, use cases, and experiences.

But anyone can use scenarios to explore future outcomes. We all make decisions based on our own personal constructs, and our mental models can be dangerous: we selectively interpret evidence, intermingle data with dubious assumptions, and listen for signals that confirm our existing biases. Scenarios invite us to dismantle those constructs as we consider alternative worldviews. They also unlock something invaluable: the ability to re-perceive reality. It can be difficult to lean into uncertainty with curiosity, rather than judgment, especially when the "What if?" questions that must be asked challenge your political, religious, or philosophical orientations. But the act of "re-perception" awakens you to the possibility of a future that differs from your current expectations. It helps you understand that you cannot know all things at all times, and that you should be curious, rather than absolutely certain, about what you perceive in the present.

In Buddhist teachings, the parable of the elephant explains the importance of re-perception. Perhaps you've heard the story before, but it's a useful reminder of the importance of re-perception: A group of blind men encountered an object. But as much as they tried to grasp what stood in their path, none of them could identify it. One man, who stood by the animal's side, thought it might be a wall. Another man, who only felt its tusk, thought it must be a hanging spear. One of the blind men jumped backward, thinking he'd found a snake. The men argued endlessly, and over time each one became more deeply entrenched in his own limited perception of reality, until there was no longer any possibility of helping them understand that the tusk, leg, and trunk were all connected to an enormous elephant.

Synthetic biology demands our re-perception. What follows in the next chapters are a series of short scenarios that describe how synthetic biology could change different facets of our lives during the next fifty years. We have considered data and evidence from synthetic biology's vast value network—that system of organizations where value is generated by all participants—as well as from academic research and investment decisions being made in different market sectors. We considered the evolution of wealth distribution and the job market, shifting attitudes toward privacy, and socioeconomic factors such as access to child care, education, health care, good nutrition, and housing. We evaluated the politics in China, the European Union, and the United States—the predominant players in the current synthetic biology ecosystem. But we also included new alliances for space initiatives, such as terraforming Mars. Synthetic biology is connected to adjacent areas in tech, including artificial intelligence, telecommunications, blockchain, consumer electronics, social media, robotics, and algorithmic surveillance, which are all playing increasingly important roles in the bioeconomy.

These scenarios should leave you with more questions than answers. Our intention is to spark debate and discussions about how synthetic biology might create better futures for all of us. Absent this public dialogue, synthetic biology will develop without public understanding, creating a dangerous asymmetry in perception. Some people argue incessantly over whether a trunk is a snake, or a tusk is a spear. But those who understand what's standing in front of them will make the decisions that impact humankind.

| TEN |

*Scenario One*

# CREATING YOUR CHILD WITH WELLSPRING

Welcome to Wellspring. Our world-renowned fertility experts and state-of-the-art assistive reproduction technologies are ready to help you create a new life. Wellspring has completed more than 3 million procedures, and our success rates rank among the highest in the nation. Every ten seconds, a Wellspring baby is born.

> ***"The genetic architects at Wellspring care deeply about their patients. They didn't overwhelm us with choices. We are confident in the upgrades we selected, and we will be forever grateful to Wellspring for helping us start our family."***
>
> —SAWYER AND KAI M..

## WHAT TO EXPECT

Every parent(s) is assigned a personal Wellspring team who will assist you throughout your reproductive journey. Your team includes a genetic architect, a digital fertility assistant, a genetic encryption specialist, a carrier liaison, a technician, and a Wellspring concierge.

For those who desire upgrades, cold storage for your embryos, or artificial incubation, your extended Wellspring team will include additional specialists.

Once your embryos have been created, your technician will conduct preimplantation screenings to ensure that your specifications have been achieved.[1] However, sometimes changes occur. During screening, we will identify single gene defects, monogenic abnormalities, and structural rearrangements and remove those embryos as candidates. Your digital fertility assistant and Wellspring concierge will meet with you to discuss risk factors for your strongest embryo candidates. You will select one (recommended) or two (if you are able to support twins) embryos for implantation in a carrier of your choice: you, your partner, a surrogate, or an artificial womb housed in our high-security incubation facility. If necessary or desired, your remaining embryos will be encrypted, frozen, and stored for use in future procedures.

## FAQ

As you begin your journey to becoming a parent, you will undoubtedly have many questions about what traits and characteristics to select during the genetic reprogramming process. Below are answers to our clients' most frequently asked questions.

### *Is the procedure to retrieve my skin cells painful?*

At worst, most patients feel a minor burning sensation. After sterilizing a small patch of skin on your forearm, a technician will inject a mild local anesthetic. Once it takes effect, the technician will gently remove a bit of your skin with a precision-tooled scalpel. Stitches usually aren't necessary, and the site typically heals within a week, without any scarring.

### *Is there a limit to how many embryos I can order?*

While our artificial intelligence systems run millions of simulations to generate the best possible genetic constructs to meet your specifications, we limit the number of embryo candidates to six. Your small but mighty group of embryos will offer variations of the traits you have selected. In our many years of research, we have found that the more options parents are given, the less satisfied they are with the outcomes. Too many choices during the genetic architecture phase can be confusing and even traumatic. This phenomenon is called the "tyranny of choice." Rest assured that our proprietary algorithms will select your desired attributes in combinations optimized for your unique circumstances.

***What traits can I select for?***

During the genetic architecture phase, your digital fertility assistant will get to know you, or you and your partner(s), individually. The digital assistant will conduct interviews to learn about your worldviews, experiences, and expectations. You will also undergo a series of genetic tests to determine heritable traits and predispositions. Once this process has been completed, we will develop a custom list of traits, and you will be invited to select from among these potentialities. They will include gender, physical attributes, cognitive dimensions, and other features.[2]

***Can I select from all possible traits, rather than the group I'm assigned?***

Unfortunately, no. There are two reasons we limit trait selection. First, your child will carry your DNA and will therefore match some of your genetic predispositions.[3] Second, certain genetic traits cannot be combined. For example, shoe size is proportional to height; if you select a height of between 6 and 6.5 feet, a shoe size between 14 and 20 will optimize your child's gait, balance, and posture. A tall child with tiny feet would face significant mobility challenges. Likewise, if you select cognitive traits such as superior analytical skill and a strong ability to memorize information, your child cannot also have superior intuition and the ability to think abstractly. At Wellspring, we strive to harmonize and balance each new life we create.

***Can I upgrade my baseline embryos?***

Certain upgrades are offered to parents who meet financial qualifications. At this time, Wellspring is pleased to offer approved upgrades for memory, body mass index, bone density, lung capacity, extended pharyngeal cavity (for improved vocal resonance), mild webbing between toes (for improved performance in water sports), and engineered hyperosmia (an ultra-enhanced sense of scent).

*Is there financial assistance available to help offset the costs of upgrades?*

Patients covered under the national health-care program are eligible for up to three IVG cycles and up to one new life creation. Wellspring's world-class assistive technology will be used to create a new life within the standard genetic norms as determined by your own, or your partner(s), screenings. For example, if your cognitive norm falls within 90 to 110 intelligence quotient points, your embryos will match your score range. Upgrades that are not covered under the national health-care program must be paid for by parent(s). Unfortunately, we do not offer financial assistance to offset the costs of upgrades.

*Do military veterans automatically qualify for upgrades?*

According to the current national government's five-year schema, all military veterans are eligible for upgrades at no out-of-pocket cost. Veterans interested in upgrades must first enroll in the Mil-Gen program. Mil-Gen program participants will be assigned an additional Wellspring liaison who will monitor the genetic architecture and screening process and select exclusive Mil-Gen upgrades for each new life. New lives created under the Mil-Gen program will be monitored until children turn eighteen, at which point they will serve their required four years in the military. After their service, they can choose a military career or transfer to an appropriate post in the government. And remember that new lives created in the Mil-Gen program are guaranteed lifetime employment and full benefits.

*How can I opt in to Wellspring's private beta upgrade programs?*

Wellspring is committed to excellence. We continually scrutinize our own protocols and methodologies to exceed the exacting standards set by the national government. As assistive reproductive technology innovators, Wellspring scientists are always hard at work developing new features and upgrades. Our beta upgrade program participants meet with a specialized team that will determine, on a case-by-case basis, whether these new features and upgrades will harmonize with the desired genetic architecture of completed embryos. Beta upgrade programs are private, opt-in opportunities between individuals and program managers. There is no charge to parents accepted into a beta upgrade program; however, the beta upgrade program should not be viewed as a workaround for parents who do not qualify for standard upgrades. ***Note:*** Mil-Gen patients are automatically entered into exclusive Mil-Gen beta upgrade programs, and such cases do not require an application.

***Which beta upgrade programs are available during my IVG cycle?***

It's always best to check with your genetic architect to learn which beta upgrade programs are compatible with your baseline genetic construct. Our current beta upgrade programs include:

- **Respiratory Upgrade:** Each lung is made up of smaller lobes, three on the right side (superior, middle, inferior) and two on the left side (superior and inferior). Fissures separate the lobes from each other, with bronchial tubes extending throughout. In this upgrade program, we include an additional lobe on each side as well as wider bronchial tubes and increase the capacity of the heart. For the athletic-minded parent(s), this upgrade is highly likely to result in improved cardio-respiratory function.
- **Improved Night Vision:** Normal eyes are capable of seeing in low light (such as moonlight or candlelight). However, we can reprogram the retina's directional neurons to send additional information to the brain. This will result in extra-perception in low-light settings, such as dark rooms and closets. In areas where there is little natural light—such as a forest or country road at nighttime—upgraded retinas will allow those treated to see most details clearly. ***Important Note:*** This beta upgrade is incompatible with a selection for blue, lavender, turquoise, green, peach, or pink iris colors, as genetic instructions to produce extra brown pigments are included to protect the retina against light exposure.

- **Thicker Skin:** *Homo neanderthalensis*—Neanderthals—produced far more keratin (a fibrous protein) than *Homo sapiens*. Compared to us, they had tougher skin, hair, and nails and were better suited for colder climates. In this pilot program, we introduce certain Neanderthal genes to generate increased keratin production. Primarily an aesthetic upgrade, it will result in smoother, fuller skin and fewer wrinkles over time, as well as generally thicker hair and stronger and longer (if desired) fingernails.[4]

***What if I cannot find a carrier for my embryo(s)?***

Wellspring's high-security, encrypted incubation facility provides a safe alternative to in vivo pregnancy. Your artificial womb compartment is customized for your genetic profile, and compartments are monitored continuously by two digital assistants and a Wellspring Incubation Specialist. Accessing your private dashboard, you can generate an unlimited number of sonograms and ultrasounds at any time. We offer a selection of sounds—including your voice(s); white, pink, blue, or brownian noise; and music—to play at optimal periods of developmental growth. On birth day, you and up to three family members enter the Opening Room to witness the opening of your compartment and removal of your baby by a team of automated delivery specialists. Your Wellspring concierge will then assist you with all of your new parent needs as you transition your new life into your home.[5,6,7]

ELEVEN

*Scenario Two*

# WHAT HAPPENED WHEN WE CANCELED AGING

As Generation Z started becoming grandparents in the late 2050s, they redefined stereotypes of what elders looked like. The skin on their hands was smooth and plump, their hair was still thick, and they were unusually agile for their age. At a molecular level, the signs typically associated with aging—genetic instability, mitochondrial damage, tissue degradation, inflammation, and wear and tear on cell membranes—were noticeably absent. All of the tiny mutations and metabolic failings that had affected every previous generation were no longer causing problems. Zoomers were getting older—but not *old*.

For such a universal condition, the *why* behind how human bodies aged had attracted relatively little attention from scientists. The basics of what constituted aging were simple and widely known: human cells could not keep dividing forever. Without cellular division, there was no growth, repair, or reproduction. Eventually, cells became senescent, acting like zombies: they were still living, but they did not function properly. Nor did they die off in a timely manner, to be flushed from the body or recycled. Senescent cells were toxic to the tissues and organs, secreting damaging morbid molecules.

Beyond that, scientists could only suggest theories about the aging process. Some believed it was related to inflammation levels and the body's eventual inability to activate the stem cells responsible for cellular repair and renewal. Others believed that aging was better understood as a failure of systems: that our endocrine and respiratory systems and microbiome degraded at different rates, causing an imbalance, until the body's machinery could no longer function properly. Still others believed that it was simply evolutionary genetics: our genomes had been shaped by a natural selection that promoted early puberty and reproduction, and after we produced offspring there was no real biological reason to persist.

Still, as some researchers had understood for decades, certain organisms had proved that the aging process wasn't entirely inevitable. Severely restricting the caloric intake of rats and other species caused them to live longer. Parabiosis research, where the bloodstream of an older mouse was connected to a younger one, showed that cellular and tissue regeneration in the older mouse could be successfully stimulated.[1] One particularly interesting experiment involved genetically engineering mice so that any individual cell would effectively commit suicide instead of going into a zombie state of senescence. The results were dramatic. At twenty-two months, which is retirement age for the average mouse, such mice still looked and acted young and healthy. Further tweaks to their genetic programming extended their lifespan by up to 42 percent.

But translating these promising results to humans was tough. Mice are physiologically very different from us. Calorie-restricted diets—a fad in the 2020s—had some effect on slowing aging. It also made people weak and tired, and even those obsessed with longevity found it difficult to stick with fasting routines. Another trendy treatment offered by wellness startups involved transfusing twentysomethings' blood and plasma into older adults. But the transfusions were expensive and risked transferring blood-borne diseases. In any event, all of these companies soon folded, because wealthy people didn't leap at the chance to pay for the privilege of being guinea pigs in risky medical experiments.

The debate seemed unresolvable even after researchers in China (at the Chinese Academy of Sciences and its Beijing Institute of Genomics, along with Peking University) completed a multiyear longevity study with one hundred thousand elderly human subjects. When that study, by far the most ambitious to date, was finally published in 2027, the results appeared to produce few solid leads.[2,3]

Or so it was thought. A few scientists seized upon some very preliminary findings about senescent cells that were buried in the report and continued the research. They began exploring ways to selectively target and deactivate such cells with small molecule drugs and virus- and nanoparticle-delivered gene therapies. By 2035, they had brought to market a new class of therapeutics called senolytics, reputed to promote healthier and longer lifespans.[4] To validate these claims for FDA approval, they had conducted extensive "rewinding" trials with these drugs often in combination with cell therapies, first in older mice, and then older dogs, and finally in people. With the old cells gone and techniques to stimulate new cell growth, the restoration of the test subjects' biomarkers to levels only seen in much younger people thrilled scientists. The most convincing evidence, though, and what so excited the general public was this: the people who'd received the treatment looked, and felt, much younger.[5,6]

One key study zeroed in on the cells in knees that secrete collagen to form the meniscus, a squishy tissue that absorbs shock during everyday activities. Eventually, adults' cells stop producing adequate amounts of collagen, the meniscus deteriorates, and the bones in the knee joint must absorb more force. As those bones rub up against each other, the joint decays, becomes brittle, and crunches against nerves. Without corrective surgery, people whose meniscuses had worn away could find the simplest movements—getting up from a chair, walking across a room—excruciating. In the rewinding trials, one injection spurred cells to start secreting collagen again, thus rejuvenating creaky knee joints. Within days, those who received the treatment started running, dancing, and playing tennis and basketball again.

Hearing loss, which occurs naturally with age, also responded well to rewinding therapies. A lifetime of loud noises, coupled with cellular aging and the loss of delicate hairs in the inner ear, once inevitably resulted in gradual hearing loss. Hearing loss has serious repercussions: it can lead to problems with walking, falls, and even cognitive decline. Even the best hearing aids, which used artificial intelligence to automatically manipulate wave forms, required people to wrap a piece of equipment around their ears. Like all hardware, hearing aids had shortcomings: they required maintenance, batteries, and upgrades. They were easily lost. But one rewinding injection could restore hearing to almost original levels in just a few weeks, instantly making all that old hearing aid machinery obsolete.

The biggest players in the cosmetics industry were quick to notice these new anti-aging technologies. They had long been looking for the next and better versions of Botox and Jeuveau, the neurotoxins long used to relax wrinkles. All began backing the development of topical CRISPR formulations: CRISPR creams, as they were commonly known, activated cells beneath the skin to restore elasticity and reduce wrinkles. Originally, they were developed to treat human papillomavirus, the most common sexually transmitted infection. The success of those first CRISPR creams, which were applied directly to the cervix, prompted researchers to investigate other use cases. It was Chinese scientists who soon discovered that a topical CRISPR gel could deactivate specific genes without causing harmful side effects. Soon CRISPR creams were available to relax frown lines, restore hair growth (or turn it off), change hair and skin color, and adjust skin microbiomes to cure acne. These latest treatments required no numbing, no needles, and no doctor, and could be precisely applied to return the skin to its natural youthful state, rather than merely freezing a brow line in place.

But these were still only cosmetic fixes. What the industry lusted for were systemic treatments for health and extended life, a market they believed was worth trillions. Quietly, they devised and put into motion a long-game strategy to make this happen.

It began with dogs, who researchers had realized decades earlier were the ideal candidate for longevity testing. Their cognitive abilities declined just as that of humans did, and their behavior and athleticism were both easier to test in the context of human environments. In the 2040s, cosmetics giants like L'Oreal Estee Lauder spent billions performing longevity research on genetically engineered Labrador Retrievers and German Shepherds. Behind the scenes, those companies also quietly commissioned major movies, novels, and op-ed essays that promoted the wonders of human longevity.[7]

It all worked. But while CRISPR proved an effective treatment for a host of age-related maladies, researchers continued searching for a whole-body solution. In the lab, a molecule called *nicotinamide adenine dinucleotide*, or NAD+, decreased in rats as they aged. It was found that the NAD+ molecules were catalysts for *sirtuins*, a set of seven genes responsible for DNA repair. Increasing NAD+ levels by 60 percent to support cellular energy and metabolism convinced cells to shift gears into repair mode, which delayed the onset of age-related diseases. There was another approach, too: Typically, cells naturally expel harmful proteins, recycle their genetic material, and extract the energy they need to survive in a process called autophagy. Under the right conditions, autophagy can also be triggered to induce programmed cell death. Researchers determined that they could target and halt autophagy as needed in order to stop cells from aging. Over time, both NAD+ and autophagy treatments, introduced in limited dosages, allowed Millennials to live much longer without immunocompromising the body or causing downstream negative effects.[8]

By 2045, Millennials and Gen Xers were consuming $150 billion worth of regenerative products and treatments each year. Most of these treatments were quite pricey, even those sold at near-cost in order to collect data or to establish new brands. But by the end of the decade Millennials and Gen Xers—or at least the wealthier, more health-minded ones—looked much as they had in their twenties, and tests showed that their bodies were blissfully free of senescent cells. That happened thanks to the new drug cocktails on the market, which had been proven in canine studies to be incredibly effective in stalling the aging process while also increasing life expectancies by almost 60 percent. Even more impressive were the gene-based therapies built on the mouse studies from decades earlier. Some dogs that had been treated with these therapeutics had doubled their expected lifespans.[9,10]

What few saw coming were the social implications of ultra-long life. The social safety nets designed to assist financially strapped and retired Americans were collapsing. It became apparent during the COVID-19 pandemic of the early 2020s that such safety-net programs were vulnerable to health crises. In the years following the pandemic, Congress failed to modernize programs meant to buffer the consequences of unemployment, loss of income, or loss of health insurance. The Supplemental Nutritional Assistance Program (or SNAP, once called Food Stamps), Supplemental Security Income, and unemployment insurance buckled in 2025, when the post-COVID-19 economy failed to bounce back. Some states and cities had local programs that helped, but lack of funding outstripped the magnitude of need. Social Security's cash flow, which had run negative for years, had long been funded through a payroll tax. But the COVID-19 crisis had marked the beginning of significant changes in how, and where, people worked. The program's sources of income, which had been deposited into a trust fund, began to dry up, and the fund, flush with cash during the Clinton era of the 1990s, was depleted by the end of the 2020s. Likewise, dozens of S&P 500 companies—including General Electric, IBM, and General Motors—now had pension liabilities nearing $2 trillion.[11] Once a perk of long-term employment for those working for the United States Postal Service, for example, or large corporations such as Coca-Cola, pensions often had to be dropped. There weren't enough new people paying into the plans, which were increasingly expensive to administer.[12,13,14]

Making matters worse were the advancements in automation that reshaped entire sectors of the economy. Labor economists had initially thought that blue-collar workers would be the first to lose employment en masse, because of the widespread deployment of self-driving vehicles, warehouse robots, and service bots designed to perform basic, repetitive tasks. But they were wrong, at least partially. They, and people like them—well-paid white-collar workers in law, insurance, and accounting—also saw their jobs swept away by automation.

As for those who still had jobs, many now expected to stay in the workforce into their nineties, and even beyond. Unionized workers whose contracts did not stipulate a mandatory retirement age leveraged their seniority as long as possible. This made the competition for jobs and career advancement downright ruthless by midcentury.

Given all this disruption, coupled with the normalization of lower-paid work-from-home jobs and gig work, Generation Z began missing key economic milestones. They were unemployed or underemployed and unable to afford travel, or to make large purchases, such as homes and cars. College graduation rates for their children, Generation Beta, dropped precipitously. Zoomers found that incomprehensible, given that their generation had been obsessed with elite schooling and pricey degrees. What was far worse, and what really angered them, though, was that all their education and effort turned out to hardly matter at all. Longevity treatments and the state of the labor markets stymied any attempts they made to advance in their careers—or even to find a decent foothold from which to start one.

Thanks to regenerative and rejuvenation therapies, older people working as teachers, in utilities maintenance or construction, and in many government roles squeezed new workers out of jobs. There were even bigger problems among the ranks of executives. The CEOs of privately held businesses refused to retire. Their counterparts at publicly traded companies also hung on, packing their boards with pals who never bothered asking when they were going to step down. Heads of family-owned companies threw out succession plans once age was considered a treatable pathology. Promises to increase gender equity and diversity among the top ranks of America's companies came to naught. The old guard refused to make space for their younger sixty-something successors, and also increasingly relied on AI-based systems to fill operational roles.

Tensions between longer life and economic opportunity weren't limited to America. Japan, which already had the longest average lifespans in the world, was now home to so-called *nijikai-jin*, which roughly translates to "afterparty people." Nijikai-jin are those who live two full lifetimes: after making it well past seventy, they plan on living, and working, for another seventy years.[15] While economic policy incentivized innovation in robotics, stemming from a stated policy to create an ecosystem that could provide automated nursing and elder care, it failed to create space for women in the workforce. Robots were "hired" to work in schools, hospitals, and Buddhist temples, and women were still expected to stay at home to cook meals, manage housekeeping, and rear children. A resistance to change resulted in women having to choose career independence or marriage and motherhood. Japan's population, which hovered around 125 million for the first twenty-five years of the century, had long been predicted to drop below 100 million. There are now 130 million Japanese people, but 86 percent of them are over the age of forty. Few children are being born. In the northern region of Tohoku, empty schools haunt the countryside.

By 2065, hostility between near-centenarians and "youngsters" in their fifties to eighties was growing, resulting in widespread protests and civic unrest. The "still young" organized support to gum up networks and stage virtual sit-ins, blocking the centenarians from working. No industry was immune. Anthony Rizzo, the first baseman who was once beloved for helping the Chicago Cubs clinch a World Series title, was now widely reviled for not retiring from playing.[16] Fans grew to rue the day they had once found so exhilarating: the day that Rizzo and his former World Series teammates Kris Bryant, Javier Báez, Kyle Schwarber, Jake Arrieta, and David Ross bought the Cubs and returned to the field. Universities deeply regretted the tradition of tenure. Hundred-something professors not only refused to update their syllabi, they also refused to retire. Geriatric news anchors and cultural critics wouldn't leave the spotlight; the same old singers and actors wouldn't leave the screen and stage. Even scientists refused to retire so that younger ones with fresh ideas could take over their labs. Scores of talented young people never had the opportunity to hone their craft or advance in their chosen professions.

In the United States, there were calls to enact a new, mandatory retirement age of seventy-five. This would have upended the Age

Discrimination in Employment Act, which Congress had passed in 1967, and which had made mandatory retirement illegal. But this new mandatory retirement age would have applied to the federal government and to Congress itself, and there was little appetite to make such changes, especially once the congressional pension funds were depleted. Republican Ted Cruz, ninety-five and looking slimmer and fitter than ever, was now entering his fifty-second year in office, still tormenting liberals as the senior senator from Texas. And he still got lots of airtime on the Fox Neural News Network (FNNN). Or at least he did when that channel wasn't having an even older senator, Rand Paul, on its shows instead. Fox viewers especially loved seeing fellow centenarians on their favorite shows.[17]

A reverse discrimination case brought by federal workers alleged that youth was now a factor in hiring decisions and job classification. The lawsuit argued that younger workers in their fifties and early sixties were not being told about promotion opportunities and their eligibility for higher salaries. Those jobs were given to people in their seventies and eighties instead, who had deep institutional knowledge and whose personal connections made them the preferred candidates. A federal district court judge ruled that workers were entitled to protection from discrimination based on age, but an appellate court overturned that decision. Now, the US Supreme Court is preparing to hear arguments that will determine whether age can be a factor in hiring. Unfortunately, pundits doubt that judges like Chief Justice John Roberts, who just turned 111, and Justice Elena Kagan, who's 105, will sympathize with the plaintiffs. Of course, those pundits are in their late eighties or older themselves, so they may have some biases, too.

|TWELVE|

*Scenario Three*

# AKIRA GOLD'S "WHERE TO EAT" 2037

What are the hottest culinary and dining trends in the greatest city on Earth? Plenty of delectable, delightful—and, dare I say, even innovative—new dishes that offer comfort and sustenance. Your wary restaurant correspondent spent six full months sampling bold flavors, time-honored standbys, and the occasional culinary experiment gone horribly wrong.[1]

On the East Side, which has long been a hotbed for Vietnamese cuisine, a cluster of upstart chefs are making functional pho. An aromatic, broth-based noodle soup, the latest functional phos are infused with synthetic diindolylmethane (we are told it detoxifies and purges excess estrogen), supercharged carotenoids (which optimize eye health and immunity), and turmeric engineered with extra curcumin (it reduces inflammation and improves focus and memory). If you're craving Sichuanese *mapo tofu*, that addictive chili-laced amalgam traditionally made from tofu and ground beef, the Lake Shore district has a new bioreactor and source supply of bovine stem cells that are used to culture meat. The intense, gamey, beefy aroma and silky layer of fat result in a dish that melts away while leaving behind a little fire and a lot of tingle. As always, there are plenty of new locations along the water that offer spectacular New Coast views that will make you feel better about eating away a sizable chunk of your paycheck.

For those who prefer custom dining experiences outside of traditional restaurants, ghost kitchens dotting the valley have ramped up their robotic staff, and the recent influx of vertical farms and culinary engineers has resulted in satisfying terrafarm-to-table options.[2] I do not miss the cramped restaurants of yesteryear, when tables were squeezed so tightly together you might accidentally spear a crouton from a neighbor's plate. Nor do I long for the days of painfully cacophonous food halls, where loud music and loud diners raised decibel levels above what even engineered people can easily tolerate. Knowing that I can reserve space anywhere that a ghost kitchen crew can set up a table and chairs, and select in advance how chatty I want my automated service to be, is a gift.

Over the winter, I treated my friends to endless gastronomic rounds of locally sourced sushi from a ghost kitchen in Bayview. I reserved one of the ghost kitchen's outdoor spaces, a pretty spot beneath a shady willow tree that shielded us from the sun.[3] Our service bot rolled up to our table, gently placed freshly cultured *nigiri* arranged on bamboo plates in front of us, and whispered a soothing "Dōzo meshiagatte kudasai" in Japanese before wheeling away. We ordered *toro* (fatty bluefin tuna belly) with additional marbling, and it arrived exactly as expected. I hosted a dinner party with two dozen friends inside the AMC Entertainment space—Remember how people once gathered in person to watch two-hour movies? It's still baffling to me—and put together a custom midsummer tasting menu from a Scandinavian Ghost Kitchen. With gorgeous digital roses, tulips, and greens lining the long tables, and mixed-reality canopies made of digital fairy lights and eucalyptus displaying overhead, we enjoyed crisp rye breads, pickled functional herring, and sweet dill from a local underground farm. To the delight of my guests, I'd advance-ordered the requisite *pressgurka* (pickled cucumbers), custom-brined to delight each person's genetic-gastronomic profile.[4]

In accordance with my annual guide tradition, I have dutifully compiled a best bioreactors list, my favorite new restaurants, ideal settings for ghost tables, and the best joints to wet your whistle. You'll undoubtedly debate me over my top picks, but know that these are the humble opinions of an overfed *bon vivant* who doesn't bristle when traditions—and the past trappings of city life—are broken.

### THE BEST NEW BIOREACTORS

We know that cellular agriculture has been booming. But bioengineers have now become really creative. They've finally recognized that the process of growing cells is similar no matter which cell types are being put into the bioreactors. So why stick with staid conventional meats? The most exciting bioreactors are culturing cells from more exotic creatures, including zebras, elephants, tigers, hummingbirds, bats, and snakes. West Side bioreactor Floria is now banking and culturing cells from thousands of different organisms. La Petite Saveur specializes in small-batch cultures. If your family prefers a bolder flavored pork chop, it can dial up the aroma and taste. But my favorites come from the geniuses at Resurrection Labs.[5,6,7]

1. ***Best Selection: Floria***

   When the founders of Floria booted up their bioreactor for the first time, they made a decent batch of ground chicken. But the team always intended to venture into exotics. After spending several years quietly building an enormous cell library, Floria opened its doors earlier this year with a world-class selection of cultured proteins. Diehard fans of Borrego—fire-roasted lamb steeped in Mexican coffee, Negra Modelo, and spices—will appreciate its soft, juicy texture. If you've never tried civets, porcupines, or bats, which I sampled at FLAB, France's first Michelin-starred bioreactor, you'll want to begin with cubed bat meat, which Floria can also grill for you on site.[8]

2. ***Best Small-Batch Meats: La Petite Saveur***
   Here's a solid option for families with picky eaters. Cultured meats can taste and feel rather bland. My kids—like yours, I suspect—demand all-vegan meals, all of the time. Vegetables are engineered for exceptional flavor and color, so I get why they couldn't care less about the chicken nuggets that you and I lived on when we were tots. La Petite Saveur brews custom orders of traditional meats—beef, pork, chicken, and lamb—to your family's exacting specifications. If you don't feel like cooking at home, the bioreactor's sister operation, La Petite Assiette, will prepare your meats with its admirable selection of handcrafted sauces and spice pairings.[9,10,11]

3. ***Best Prehistoric Meats: Resurrection Labs***
   I remember the first time I took a bite of a woolly mammoth steak. The meat was tangy and minerally, like bison, but had a hearty and slightly sweeter overtone. It was also slightly gelatinous, because the mammoth meat had been tenderized to make its otherwise rubbery and coarse texture more pleasing. (It didn't, but points for trying.) The meal came courtesy of an omnivore-themed party I attended with some well-to-do friends in Sun Valley, who had hired synthetic biologists to culture genomes of extinct animals. So when Resurrection Labs opened in the Arts District, I was apprehensive. The first few mammoth steaks I ordered were epic failures—the first smelled like a urinal, and the second was too tough to chew. I gave them a few months to work out the peculiarities of their bioreactor, during which time they wisely replaced their chief bio officer, who clearly was more interested in science than succulence. Now, Resurrection is culturing a small, refined menu of extinct species, including Pyrenean ibex, passenger pigeon, dodo—and, I'm happy to report, its extraordinarily toothsome woolly mammoth.[12]

TRENDS WE'RE TIRED OF

1. ***Confused Service Bots***
   Some days I miss human service staff. Yes, they were often too chatty, too forgetful, or too slow. But they could at least empathize with us needy diners. Service bots, acquired on the cheap, mistake our calves for table legs, misjudge the distance between their robotic arms and our tables, and often have no idea what we're saying. Two weeks ago, a bot in the process of delivering a steaming hot plate of sinus-clearing green chutney, creamy vindaloo chicken, and airy naan, extended its arm toward us and attempted to drop our food from a good five inches above our table. My companion slapped its emergency stop button just in the nick of time.[13,14]

2. ***AI Transparency Labels***
   Lines and lines of code do not make for a lovely *amuse bouche*. We don't need to know your algorithm's back story, or who built your in-house database. Data is a downer, so just stick to the government regulations, keep your labels available for those who have the fortitude to read them through, and trust that your diners are comfortable with whatever AI systems and sequencers you've used to genetically design, culture, and grow your ingredients.[15]

3. ***Pre-Menuing***
   Just because you have access to our daily metabolic rates, our food and drink preferences, and our activity history doesn't mean we've lost all agency in meal selection. For the love of Julia Child, please let us pick what we want to eat! Allow us the courtesy of selecting from a wide variety of dishes, instead of extrapolating from our biome markers and metabolic scores! We may be in ketoses and burning optimal levels of energy, but we didn't come to your restaurant to stay that way.

4. *Hipster Cocktails*
   Why, dear readers, are we celebrating the worst excesses of 1990s bar culture? Molecular cosmopolitans should be in no bartender's repertoire. And yet—ugh—they are everywhere. Just because you're using the genetic sequence of *Microcitrus australasica* (the once super-rare variety of Australian finger limes, for the uninitiated) doesn't make serving this tired, old, too-sweet cranberry juice cocktail acceptable.[16]

5. *Microdosing Mushrooms*
   I love our noble psychedelic mushrooms. Who among us doesn't appreciate how they unlock creativity and imagination? But no one needs psychedelics on every single menu around town. Especially since most of us already microdosed in the morning.

## MY FAVORITE SETTINGS FOR GHOST TABLES

One welcome addition to our city's dozens of ghost kitchens: ghost tables. Who could have imagined that the initial success of Airbnb would someday lead to fractional land ownership? There are now 1,260 spaces available in our town for hourly rent, all accessible by delivery bots. Reserve your spot, and the restaurant now comes to you. Not all locations are ideal, or even safe. The Geok Ghost Kitchen, purveyor of California-fresh-meets-Cambodian-street-food fare, had an opening on its "beautiful veranda with skyline views." I pre-bought three hours and figured we would order our meal once we arrived, but the space was a dilapidated mess. One of my companions stepped onto what we thought was a secure step. Then her foot went through it, sending splinters and wood shards everywhere. But on another evening, we rented a table inside a new studio space that projects mixed-reality art onto blank walls. It was springtime in Japan, and we selected a cherry-blossom themed digital installation to complement our meal. You'll want to investigate the city's varied ghost tables for yourself, but these new finds are worth a trip.

*Rooftop Deck of the Artep Hotel*

Perched high above the city on the eighty-seventh floor is the Artep's ghost table collection. The panoramic skyline views are unparalleled—you'll see the Old Coast and the New Coast, and the virtual and invisible paneling means that you will enjoy the quiet solace without either street-level noise or the momentary terror of being blown to and fro by wind gusts. Make reservations several weeks in advance and plan on a second stop for dessert: Artep only allows two-hour seatings.

*Forest Glen*

The lush greenery of Forest Glen seems oddly juxtaposed with our loud, bright, bustling city. The technology originally used for that venerable first-generation carbon-capture project couldn't be upgraded. Fine by us, because we now have a splendid private park to enjoy. You can rent tables for long blocks of time and order service from several nearby ghost kitchens. Given the natural theme, we recommend Mozaic's vegan menu.

*Bella's Basement*

This quirky former comic book store is now a lively spot for an easy, casual meal with friends. At the height of Marvel movie madness—when it seemed like a new Black Panther movie was premiering every summer, and minor characters got their own meandering shows—Bella's Basement was the place to find action figures, memorabilia, and, of course, comic books. Much of the original décor is still intact: Captain America's shield is bolted on the ceiling, an enormous mural of Morpheus from the *Sandman* covers the wall, and assorted toys are squeezed into every available nook and cranny. If you haven't been—and don't worry, this isn't a spoiler—do not order your meal in advance, or you'll miss a clever opportunity to meet Jarvis.[17]

## BEST JOINTS TO WET YOUR WHISTLE

As you know, ever since Day Zero, when our water taps started to go dry, desalination systems have given us all of our drinks. While reverse osmosis systems are mostly used throughout the city, some of the city's best bars are instead starting to use halophytic algae-based processors. These employ organisms to remove salts from brackish water and also remove CO2, and the residue is later dried and made into useful feedstocks for animals. With all the new molecules available, mixologists are serving depression-fighting beers, libido-enhancing elixirs, and some of the smoothest molecular whiskey you've ever had. My companions and I usually start our night off with a hangover-killer enzyme elixir made at Zingoff's on Ninth Avenue, which magically and mercifully omits the metallic aftertaste common in mass-market brands.[18,19,20]

## DRINKS DURING WORK

Hidden in a subterranean space beneath the McHarron Hotel in Station East is a cozy hideaway that's only open during regular business hours. (Last call is at 6 p.m.) The playful cocktail menu, by bartender Emma Harper, includes her signature "5-Minute Break," a home-brewed molecular whiskey mixed with ginger shrub bitters, crushed recycled wastewater ice, and a sliver of dehydrated orange. Our favorite is the WFHFW (Working From Home From Work)—Harper's go-to drink for those harried days of meetings that blend together. She mixes dark rum engineered from synthetic sugar cane with molecular Cuban coffee–based liqueur, a dash of simple syrup, and two dashes of synthesized chocolate bitters. You'll catch a quick buzz—and stay awake once you're back at your virtual office.

### *Spritz and Fitz North*

The team behind the beloved Near Shore dive bar Spritz and Fitz opened a second location farther up the avenue. But don't let a nicer and more picturesque neighborhood fool you into thinking this new outpost is overly fancy. You'll find the same engineered grime on the walls, the same custom-brewed microbes, and the same selection of more than two hundred nano-brews on tap. As expected, a hologram of Fitz Larson, progenitor of the bar's original yeast strains, hangs proudly over the bar.

*Château Gact*

Ever since Elijah Codding opened Château Gact in 2028, locals have flocked to his weekly tasting menu of custom microbial wines. His sprawling, underground vineyard got its genetic source code from black Tempranillo and Sangiovese grapes, and it produces two noteworthy wines. Its Reserva Especial is a small-batch, medium-bodied synthetic red—refreshing, not too mellow, and perfect for a midday meal. Splurge on his peppery, citrusy Domaine de la Ásvestos, which Codding created using custom molecules to synthesize the lime-scented, volcanic soil of Santorini for his synthetic Assyrtiko varietal. Pour me another glass![21,22,23]

THIRTEEN

## Scenario Four

# THE UNDERGROUND

Farmers in Mymensingh, a northern district in Bangladesh, had been cultivating seasonal crops, including mustard, jute, and rice, for generations. But in 2030, as sea levels rose and unprecedented floods washed away acres of land, the government struck a deal with China. As part of China's Belt and Road Initiative, the Chinese Communist Party's massive infrastructure project launched a decade earlier, the CCP offered to help Bangladesh deal with this climate catastrophe. Its plan was to reroute water to create artificial islands in the Bay of Bengal and construct high-tech seawalls to keep the seawater out. The deal also promised engineered strains of saltwater-resistant rice, which—thanks to China's dominance of genetic editing technology—would produce viable crops and rice fit for human consumption.[1]

But the rice never materialized, and by 2035 it was clear that a few seawalls wouldn't save the country from newly extreme seasonal fluctuations in water levels brought by rising tides and increasing monsoons. Bangladesh had lost 18 percent of its land to rising sea levels, which forced 15 million people living in low-lying coastal regions to relocate. As flooding worsened, septic systems failed and drinking-water facilities were fouled. People lost their homes and their livelihoods. Traditional rice paddies, inundated with seawater, could not survive. Farmers, who for generations had cultivated vast wheat, maize, and potato fields, and who were skilled in working around seasonal flooding, tried to move inland and as far north as possible. But increasing competition for jobs and housing left them with few options, and many stopped farming. Some tried to cross the borders into Myanmar and India, claiming climate refugee status, but with millions of climate refugees in those countries already scrambling to survive, both countries refused them entry.[2]

As extreme weather events mounted—including daytime temperatures topping 63°C (145°F) in Saudi Arabia and days of more than 30°C (100°F) in Siberia—crop production in dozens of countries failed. Drought, followed by sudden floods and landslides, made it impossible to grow wheat in Afghanistan. In South Sudan, where peanuts, sesame, sugarcane, and millet once grew easily, daytime temperatures routinely exceeded 49°C (120°F). Massive dust plumes stretched across the country and into Ethiopia and Kenya. The commercial rubber farms in Liberia, which had operated some of the largest natural rubber production facilities in the world, no longer met capacity for exports.[3,4]

Meanwhile, America's breadbasket, in its Great Plains, where the majority of grain and corn had long been grown, migrated far north. For a time, the best location to cultivate cereal grains was the upper Great Lakes regions: the northern parts of Minnesota, Wisconsin, Michigan, and New York, which used engineered soil to accommodate these new crops. But as global temperatures continued to rise, the region didn't just become warmer; it grew more atmospherically unstable, which resulted in firenados (swirls of wildfires and tornados) and *derechos* (straight-line windstorms that bring hurricane-force winds and torrential rains).[5]

In November 2036 the United Nations Climate Change Conference released a disturbing scientific study: with almost nine billion people, humanity was running out of room to grow. Allowing urban sprawl to continue would threaten food production and cause the collapse of the already overstressed ecosystems. There were now only two options for continued human expansion: to start building down into the Earth, or to leave the planet behind entirely.[6]

For decades, Elon Musk, CEO of Tesla and SpaceX, had been insisting that humanity's best shot at long-term survival was to eventually become a multiplanetary species. He cited elevated levels of carbon pooling in the Earth's atmosphere, extreme droughts, and the loss of biodiversity as precursors to a looming catastrophe. He began development on a program called Starship in 2016, which was intended to ferry cargo and, eventually, one hundred passengers between the Earth, Moon, and Mars. By 2021, NASA contracted SpaceX to develop a modified Starship vehicle for its Artemis program. Musk focused on building the core infrastructure that would eventually be required to sustain life, whether on Earth or on the Moon, Mars, or even beyond. But Musk realized he could not build an off-planet living environment on his own. Ever the showman, and with his personal fortune approaching $1 trillion, Musk announced an audacious contest called the Colony Prize. He'd award $1 billion to any team that could build and operate an underground, airtight colony of one hundred people for two years. In other words, the ultimate Mars simulation.[7,8,9]

Musk knew that for humans to thrive off-planet, regenerative systems would need to be developed at a scale never before achieved.[10] The International Space Station once housed as many as thirteen astronauts on board, but typically only six or seven lived in the ISS at once. Colonists also needed to grapple with long periods of confinement. The typical ISS mission was about six months.[11] NASA astronaut Scott Kelly had spent nearly a year in space. Cosmonaut Valeri Polyakov held the record for a single mission, spending an impressive 437 days on the Mir station in the 1990s.[12] A better example for understanding a hundred-person enclosed society was a submarine—but even here the longest submerged and unsupported mission topped out at 111 days.[13] Winning the Colony Prize would require keeping the doors sealed tight for more than 700 days.

The contest rules were simple by design. Entrants were to outfit and assemble airtight canisters into a closed living environment. Those canisters, which would begin as empty, modular self-contained spaces that would fit into the cargo bay of a rocket, could be configured as living quarters, science labs, farms, schools, water treatment systems, manufacturing facilities, and anything considered necessary to support a community. Colonies were encouraged to include facilities to support concerts, sports, and other forms of recreation.[14,15] Once configured and loaded with supplies, the doors would be sealed and the mission clock started. The goal wasn't to reinvent Buckminster Fuller's geodesic dome. It was to invent entirely new networks of modular structures—using something akin to the Minneapolis Skyway System, the world's largest contiguous system of enclosed structures and bridges—that could scale to eventually become a city. Over time, such a plan would replicate some aspects of life in the days before extreme weather became our new normal.

In addition to container reconfiguration plans and simulations, contest entrants were told to submit a list of potential colony inhabitants, justification for their selection, and a detailed plan to ensure quality of life. There was an important caveat: the colony couldn't just be made up of a bunch of carefree early twenty-somethings who, back in the halcyon days of the aughts, might have attended Coachella. Every colony had to mirror the full spectrum of society: a mix of families, couples without children, and single people. The prize was intended, in part, to test population expansion in a closed system. That meant that facilities to manage pregnancy, delivery, and infant care, as well as various health issues and the various stages of life, had to be built.[16]

There were no requirements or quotas to ensure diversity of thought, ideology, race, ethnicity, nationality, or culture. Nor were there stipulations to prevent certain people from being excluded from a colony. If a group could prove in simulation that its plan could support life for two years, and if they could explain how inhabitants would work, attend school, receive medical attention, cultivate resources, and maintain balance within the colony, they'd be eligible to advance.

Selected teams would have eleven years to build, refine, and live within their structures. In the event of a system failure or need to make major changes to a configuration, any colony was allowed to reset the clock and start again, provided they had enough time remaining within the eleven-year limit.[17,18,19] There was no limit to how many colonies could win the $1 billion award on successful completion.

Colonies would receive support from Musk's various companies—SpaceX, Tesla, The Boring Company (his tunnel and underground infrastructure company), Chia (the energy-efficient blockchain and smart transaction platform), NovoFarm (an indoor, precision agriculture company), Neuralink (the implantable brain-machine interface company), and Programmable Matter (maker of materials that can shape-shift to respond to the environment or user input).[*,20,21,22] Feasibility studies and infrastructure pathfinders had been completed, so a crucial aspect was location: future Mars colonies would need to be built underground. Mars lacks a magnetic field, and radiation levels on the surface are dangerously high. The surface is cold. Building underground would provide radiation shielding and thermal insulation.[23,24,25]

Tunnels would be made by The Boring Company. Its automated Prufrock V machines could "porpoise," meaning they could be launched from the surface, tunnel underground at a rate of almost one mile a day, and then surface after completion. Tesla produced stainless steel cylinders that fit neatly into these tunnels. They were like shipping containers, except that they were shaped like canisters of Pringles chips, and had electric drive systems so they could move slowly under their own power. The interior of the canisters could be customized to house virtually anything, such as private quarters, hydroponic farms, or surgical facilities. They could operate independently for a short period, but they would typically be linked together to form more complex systems, the most straightforward configuration being a chain, like a subway train. Tesla also built solar and battery systems, while SpaceX handled transportation and communications with its Starlink satellites. The companies had

---

* We are hypothesizing that Elon Musk in the future may own or create these companies.

installed systems on the moon under its contract with NASA. Colonies would have plenty of electricity and bandwidth.

Colony Prize teams were allowed to use this research to support their designs. Digital plans, models, and specifications were available online, and empty canisters were available for purchase from Tesla for $250,000 each. The big challenge for teams would be putting together, populating, and operating complete systems.

Musk made it clear, in his instructions to those vying for the Colony Prize, that ambition would be rewarded:

> *The goal is to create habitable places not just to live but live well. Build the colony where you and your family can thrive, with the right kind of people, and consider how it might continue to grow to become fully self-sustaining.*

Colonists were required to self-fund their plans, including paying the salaries of those developing the colony and salaries for the colonists themselves. The $1 billion prize for successful colonies would be used to reimburse investors, pay out bonuses, and possibly fund further expansion. Musk believed that this model would incentivize and promote cooperation among various colonies, which would create a flywheel for innovation and accelerate a space-based economy, plus provide real-world experience on governance and operations.

The sheer magnitude of the prize—along with the weather on the surface, which had become extremely unpleasant—catalyzed an enormous global investment in closed living system R&D. The only example that came close was by then fifty years old: the Biosphere 2 in Oracle, Arizona, which completed construction in 1991.[26] Originally intended to demonstrate the viability of closed ecological systems, Biosphere 2 was ultimately plagued with problems. Too little food, poor oxygen circulation, and a power struggle over the project's management and administration doomed the experiment. No one since had tried to integrate the tremendous progress that had happened since then in vertical farming, manufacturing, sensor systems, and biotech into another closed system.

While there were tens of thousands of applications, only 180 proposals passed the initial rounds of filtering. They came from North America, Western Europe, United Korea, China, and India in what became known as colony-forming units, or CFUs. Getting started required that CFUs produce detailed plans and models for water regeneration, biofoundries, medical treatment, oxygen generation, and carbon capture. Doing so required real ingenuity and extensive computer-aided modeling. Ultimately, seventy-two CFUs had built skilled teams and secured land for their colony and surface operations. All of them had secured enough funding—ranging from government grants to investment from private companies to checks written by wealthy donors—to begin building.

Tesla began shipping thousands of canisters and power systems to colonies, which included such varied locales as Bloomington, Indiana, and Humboldt, Iowa; Dalmeny, Saskatchewan, and Edmonton, Alberta (Canada); Hwaseong (United Korea); Beizhen and Dadongzhen (China); Harda (India); Rumuruti (Kenya); and Knutsho (Norway). Along with partners, teams began customizing canisters and linking them together to form what, from a distance, could be mistaken for high-tech hamster enclosures, setting them up on the surface first and testing them extensively in preparation for moving underground.

While the prize rules did not limit the number of awards, it did set strict performance milestones. To meet milestones, colonies had to engineer microbes, including bacteria, that enabled crops to fertilize themselves. Sustainable indoor farms, which included climate-controlled environments, cloud-based AI systems, agricultural sensors, and collaborative robotics, were required to prove they could maintain safe levels of nutrition, carbon dioxide, oxygen, and hydration. Teams also had to design, build, test, and deploy DIY vaccines and therapies to manage any novel pathogen that might arise in the enclosed environment. Auxiliary products intended for everyday use—such as intelligent packaging made of polymers that effectively self-destruct or "unzip" when exposed to light, heat, or acid—were necessary to meet stringent waste management criteria.

At first, the teams struggled to meet their milestones. Creating a sustainable canister for one family to survive for a few years was hard enough. Scaling it to an entire community, and having some semblance of normal life deep underground was a far more complicated endeavor. Colony teams quickly realized the best strategy was cooperation, since there was no cap to the number of winners. Once they started sharing what they'd learned, the engineering of key colony systems evolved incredibly quickly. It didn't take long for the teams to arrive at configurations that computer simulations predicted would support 100 people, then 150, then almost 200. They all also realized that it was important to build in some redundancy. Things go wrong, such as equipment failures. And sometimes things go right. Colony populations were expected to grow during the mission experiment.

By January 2043, just six years into the experiment, the first colony, Endeavor Sub Terra, announced that it was ready to seal the doors and start the mission clock. Endeavor Sub Terra's community (ESTers, as they became known) was situated just east of the Arizona State University campus, beyond the Maricopa First Nations Community. (Ironically enough, Biosphere 2 had stood nearby the Arizona State campus, too.) It was sponsored in part by the university and the state government, which provided land and generous tax incentives. ESTers were carefully selected from the large community that had formed to build the Arizona colony. Many were families with kids, though there were young couples and people in various other relationship configurations. They'd all been living and working in the canisters for some time already. Going on the clock just meant not going outside for more than 730 days.

Its canisters were moved underground; the tunnel was sealed and filled with gases that closely matched the Martian atmosphere composition. Power and communication systems simulated the expected kilowatts and also transmission delays, which ranged from three minutes to as long as twenty-two minutes, depending on the relative positions of the planets.

ESTers were the first colonists to seal themselves off from the surface, but because of the extensive information and infrastructure sharing, most of the others were close behind. By the spring of 2044, all seventy-two colonist teams had moved underground.

CFUs devised and used different economic and governance systems. Some paid colonists as full-time employees, who earned salaries for the time they spent working on the prize and living in the community. Like the International Space Station, there was nothing to buy or sell. Salaries earned were deposited into colonists' bank accounts for their use back on the surface. Other colonies developed universal basic income (UBI) models, where all inhabitants received a set of credits to start with in the form of community digital tokens. Gradually, community members would use these tokens as currency—to pay for goods and services while they lived in the colony.[27]

There were detractors. Some people referred to the colonies as "ant farms," "hamster cages," and "self-filling prisons." But the colonists shrugged off the barbs. They believed their canisters and tunnels were great places to live, work, and raise a family. The environment was free from pathogens. Extreme surface weather events weren't even noticed underground. Tunnels proved safe during the firenados that were ravaging large swaths of North America and Western Europe in the summer of 2044.

The colonies excelled at bioengineering. Their life science canisters were equipped with the best biofabs, including sequencers and synthesizers. Those in charge of developing organisms needed for vertical farms and recycling systems invented novel approaches, and they adapted and evolved their local, natural ecosystems as time passed. They also designed special surveillance systems to detect any contaminations or mutations.

The underground colonies provided refuge from perilous surface storms, but the experiment didn't change basic human nature. Before sealing, psychographic data were collected on all community members to ensure they could withstand living in an enclosure with just ninety-nine other people, but no one accurately predicted the ideal community composition. Neurodiverse candidates were allowed, though people with panic disorder or attention-deficit/hyperactivity disorder (ADHD), or who were prone to depression, were strongly discouraged. Those who had anger management issues, or who displayed signs of narcissistic personality disorder, were typically excluded. Still, some community leaders bent the rules, or outright broke them. Wealthy donors expected access and privileges in return for their investment, which often meant jumping the line in front of more qualified or suitable candidates. Some donors even bought their teenagers a colony stint, hoping it would land their kids at more prestigious colleges later. Others thought it would be the ultimate status vacation, or a way to boost traffic to their virtual media channels, and insisted they make the list.

There were also failures. In some colonies, vicious politicking, infighting, and scandal plagued inhabitants the minute the doors closed. In Visionary Valley, for example, the funders were determined to manage the community as they would a business. Within two months, the colony imploded. The funders had insisted that only they should know the lock codes to key resources, such as food and water stores. They also built a colony-wide surveillance system that could be viewed using only their own biometric authentication. This wasn't known to colony members in advance, who realized, once they were all underground, that a hierarchical system was in place that mirrored the power and wealth imbalance they'd endured on the surface. The colonists attempted a coup, but they were effectively living in a panopticon, and there was no way for them to take over. Disgusted and enraged, they broke the seal on Visionary Valley and vowed never to return.

In every community, some colonists struggled with social isolation, the abrupt change of lifestyle, and restricted movement. Several felt a sense of persistent unease, which led to difficulty concentrating and sleeping. For others, depression and anxiety were more acute. Those colonists became easily frightened and developed paranoia. Some had violent outbursts or became detached from family members and friends. Colonists gave that condition a name—traumatic below-surface syndrome, or TBSS—and there was no easy way to treat it.

The most successful colonies were those that acknowledged humanity's basic physiological and safety needs. People wanted to feel a sense of purpose and belonging, and there were plenty of jobs to do within each community. A few UBI programs were successful, but most digital token systems weren't perfect. Colonists blew through their initial allotment fast, and there was no bank to lend them additional credits. They had to borrow from neighbors, which caused friction, as it always has. In one colony, a sudden surge in demand for strawberries led to inflation, temporarily causing the prices of all produce to spike.

Flat power structures rarely work; some people will always want to lead, and others never do. Many colonies developed a modified social democratic system of government favoring consensus. Colony administrators rotated through positions, which wasn't always perfect, but incentivized administrators against leaving a mess for their successors. Several colonies experimented with letting AI systems run everything.

Endeavor Sub Terra, the first to go on the clock, was also the first to win the $1 billion award in early 2045. Musk and the Colony Prize would eventually award fifty-five of the seventy-two teams. He considered it to be the best return on investment he'd ever made. Humanity had built the technical and social foundation for becoming a multiplanetary, spacefaring species—one that could scale indefinitely when given access to energy and raw materials. In addition to producing net surpluses of food, water, and other necessities, many colonies had reached economic escape velocity: the research, systems, and products they created were earning them a lot of money on the surface. If they wanted to, they could reinvest and keep growing. Which is why many ESTers decided to remain underground even after the mission ended.

They had developed an airlock and decontamination system that would enable colonists to pop back up to the surface on occasion to visit old friends or to enjoy one of the few days of good weather. They agreed to wearing or ingesting sensors, colony-wide testing, and quarantines to ensure that no one brought a virus or other pathogen back into public areas in the enclosure. They purchased their own tunneling machines and additional canisters to accommodate another two thousand people—but they already had a third growth plan underway for millions of colonists, with new underground neighborhoods, geothermal generators, massive bioreactors, and even an underground ocean. It may not have been his intention, but Musk's Colony Prize had seeded the largest investment in sustainable communities that humanity had ever seen.

Around the world, surface ecosystems were rewilding as struggling farms and towns were abandoned for the underground. Buildings, roads, and homes were left to degrade naturally, eroded by sunlight, water, and vegetation. Nature and natural systems were bouncing back faster than anyone predicted, which required a new generation of naturalists and ecologists to study Earth's dramatic new ecosystem shifts. For the first time in over a century, the CO2 levels in the atmosphere began to drop.

ESTers could see a future of flexible living: a way for people to live well on spaceship Earth or, if desired, off-planet. A personal module could be shipped to Mars and connected to a colony.

Sometimes ESTers would visit the surface at night. Laying on the ground, and without any light pollution, they would marvel at the canopy of stars above them. The stars seemed to whisper as they twinkled: *Come, humans, explore!*

Mars, and other planets, were waiting.

| FOURTEEN |

*Scenario Five*

# THE MEMO

Federal Bureau of Investigations

San Francisco Field Office

OCTOBER 11, 2026

MEMORANDUM FOR: FBI Director

SUBJECT: Emergency Assistance Requested to Combat Novel Cyber-Biological Attack

On October 9, 2026, at 5:23 p.m., our San Francisco Field Office responded to a mass casualty incident at the 23xGenomics campus. When FBI agents arrived, all eight laboratory personnel were unresponsive, and all had blood streaming from their eyes, noses, ears, and mouths. Private security hired by 23xGenomics reported a chemical accident; however, FBI field agents did not observe any chemicals present. Agents collected samples for investigation and secured the laboratory.

On October 10, the San Francisco Field Office received an anonymous call alerting agents to a 4chan board post from the Dark Chaos Syndicate claiming responsibility for the incident. The Dark Chaos Syndicate, which was previously known to the FBI, is a decentralized, militant, anti-GMO collective with members in several countries, including the United Kingdom, Russia, Germany, Sweden, Brazil, France, India, Iceland, and the United States. Syndicate members spread conspiracy theories using end-to-end, encrypted communications chat apps like Telegram and Signal, which continue to be difficult to monitor.

We uncovered a message board on www.gag.org in which members discussed various conspiracy theories related to genetic engineering. Syndicate members believe that the CIA developed the COVID-19 vaccine during the height of the Black Lives Matter movement to make people more submissive, and that Walmart, CVS, and Johnson & Johnson are secret branches of the government that have worked together to coerce Americans to get the shot. Syndicate members think that the vaccine enters the nucleus of a cell and that it alters DNA permanently in a way that makes humans far more placid. If people are not biologically capable of getting angry, the Syndicate's discussions contend, then they will stop protesting and obey law enforcement agencies. Agents uncovered chatter calling for action against genetics companies dating back to June 2021.

At this time, we believe that the incident at the 23xGenomics laboratory was not an accident, but rather a targeted hybrid cyber-biological adversarial attack involving the laboratory's computers, a genomic synthesis company in China, and a private-sector supply chain. This appears to be a new kind of attack, one that combines traditional cyber-hacking and genetic engineering into a new and lethal form of bioterrorism.

Based on the Dark Chaos Syndicate's posts, a nationwide cyber-bio attack on other compromised facilities could be underway, which would cause cascading failures of multiple forms of critical, life-sustaining infrastructure.

## BACKGROUND

The company 23xGenomics researches agrochemical and agricultural biotechnology with a focus on gene editing and applied technologies. The bioengineers at the lab were working on a project to develop synthetic strains of vanilla. The Dark Chaos Syndicate does not seem to be particularly concerned about vanilla, or about GMO vanilla. Nor was 23xGenomics doing groundbreaking genetic research. We believe that 23xGenomics was targeted because it was cutting some corners to obtain genetic material from China that is cheaper and more quickly produced than is possible in the United States, and that this made it vulnerable to hackers.

The physical and digital infrastructure that 23xGenomics used to support its synthetic biology work—data, DNA and other genetic materials, laboratory equipment, communications networks, supply chains, and personnel—was compromised, resulting in an unprecedented malware attack.

## ATTACK DETAIL

23xGenomics planned to develop a synthetic strain of vanilla that could be created in a lab using few resources. Researchers were developing experiments to test the vanilla's hardiness under different conditions. A previously unknown attack vector within 23xGenomics allowed Dark Chaos to enter a commercial operating system for the purposes of monitoring, performing data exfiltration, and planting malware. This is our current assessment of the attack:

1. A bioengineer working at 23xGenomics downloaded a compromised browser plug-in designed to automate the Synthetic Biology Open Language (SBOL) for data submission to online repositories. This plug-in was not blocked by the 23xGenomics IT department and it facilitated a man-in-the-middle attack.

2. The bioengineer and his team designed an experiment using the company's data sequencing software. The usual simulations were run to detect anomalies and to validate the sequence.

3. The bioengineer then ordered synthetic DNA from Livivo, a Chinese vendor that 23xGenomics uses for all of its genetic materials, enrichment panels, and kits. Livivo was chosen for its low cost and because it works faster than US companies, which, unlike Livivo, follow all International Gene Synthesis Consortium screening protocols. 23xGenomics exploited a US Department of Health and Human Services exemption for particular sequences of concern, which allowed them to circumvent certain screening protocols.

4. Between 23xGenomics and Livivo, malware was used to obfuscate the genetic sequence. The malware replaced the DNA sequence with malicious biological code, and did so in such a way that screening software could not detect that the sequence had been changed.

5. Livivo produced the synthetic DNA order and shipped it back to 23xGenomics. The bioengineer and his team sequenced the DNA using the compromised computer system at the 23xGenomics laboratory.

6. When the team of bioengineers continued their experiment, they used the malicious DNA, which they combined with other material. Rather than running a routine experiment, they unknowingly created and unleashed a lethal pathogen.

7. Multiple weaknesses in the DNA supply chain, including software, biosecurity screening, and end-to-end protocols, were responsible for this cyber-bio attack.

PROBABLE DEADLY OUTBREAK

The 23xGenomics laboratory received its order from Livivo on October 5, and according to lab records it was used the morning of October

6, giving the pathogen a 72-hour incubation period. In the three days after exposure, the eight lab employees came into contact with an estimated 120 people. Depending on the transmissibility of the pathogen, an exponential catastrophe could be unfolding right now.

The San Francisco Field Office has been in contact with the San Francisco Department of Public Health and the Centers for Disease Control and Prevention, which is currently investigating the pathogen to determine its genetic sequence and what, exactly, it is. An autopsy report showed that one victim's arteries, veins, and capillaries were leaking blood and plasma. The forensic pathologist we interviewed described "organs that completely liquefied" and "cells that look like they exploded spontaneously."

REQUESTED ACTIONS

Protocols for biocontainment should be enacted by federal, state, and local authorities immediately. Additional measures should include:

- All labs that received DNA or other genetic samples in the past five days should be sealed and closed. They could be contaminated and dangerous.
- All other research labs, commercial enterprises, and government agencies involved in any aspect of synthetic biology should immediately shut down and unplug all computers, sequencers, assemblers, and any other equipment.
- Information security officers and IT managers should identify and remove all threat actor-controlled plug-ins, software, and accounts and identify persistence mechanisms using remote access.
- Travel should be restricted or halted entirely. California does not have standardized contact tracing measures in place. There is no way of knowing how many people might have already traveled out of the city to different destinations around the state or to other states or countries.
- Shelter-in-place orders must go into effect, at the very least in San Francisco, and possibly elsewhere as well. Emergency protocols are required to enforce shelter-in-place mandates.

REQUESTING ASSISTANCE

Agents in the San Francisco Field Office contacted several agencies for guidance. These were the responses we received:

- NATIONAL SECURITY COUNCIL: Said it could begin looking into the cyber-hack, but that the CDC, HHS, and NIH would need to be brought in to assist. Referred us to the BioWatch Program of the US Department of Homeland Security.
- DEPARTMENT OF HOMELAND SECURITY BIOWATCH PROGRAM: Informed us that the BioWatch Program only provides risk assessment for traditional biological attacks through the DHS Countering Weapons of Mass Destruction Office, and that it has no authority in this matter. Referred us to DHS S&T Directorate.
- DEPARTMENT OF HOMELAND SECURITY SCIENCE AND TECHNOLOGY (S&T) DIRECTORATE: Informed us that it focuses on risk-based chemical and biological countermeasures. It has no authority in cybersecurity. Referred us to the Cybersecurity and Infrastructure Security Agency (CISA).
- CYBERSECURITY AND INFRASTRUCTURE SECURITY AGENCY: Said that it could mobilize support to investigate the malware attack, but that it did not have specific expertise in genetic code. We await their update.
- CENTERS FOR DISEASE CONTROL AND PREVENTION: We contacted the CDC to alert them to the possibility that a new virus or other pathogen could currently be spreading. The CDC is now coordinating an investigation into what the pathogen is but informed us that it does not deal directly with cybersecurity. Because we believe other labs could be compromised, the CDC recommended that we contact the National Security Agency or the Department of Defense.
- NATIONAL SECURITY AGENCY: Told us to review "National Security Council Memo, Section 5" and referred us back to CDC.
- DEPARTMENT OF DEFENSE: We did reach the Joint Program Executive Office for Chemical, Biological, Radiological and

Nuclear Defense (JPEO-CBRND), which is a part of the DoD's Chemical and Biological Defense Program, managing investments in chemical, biological, radiological, and nuclear defense equipment and medical countermeasures. JPEO-CBRND protects the Joint Forces (Army, Navy, Air Force, Marines, Coast Guard, and First Responders) from weapons of mass destruction. We were told that JPEO-CBRND would not intervene unless military assets or government property were attacked. Since the lab was private property, JPEO-CBRND would not get involved. DoD recommended we speak with the DoE.

- DEPARTMENT OF ENERGY: We were informed that the DoE's Genomic Science Program's mandate covers the research and redevelopment of biofuels, but unless the attack compromised the US nuclear stockpile they could not assist.
- FEDERAL EMERGENCY MANAGEMENT AGENCY: Finally, we contacted FEMA, primarily to alert them about a possible deadly attack that could lead to the deaths of untold Americans. We were assured that there was a National Response Framework in place. If other labs become compromised, we were informed that the Framework covers cascading failures resulting from natural disasters and other emergencies. When we asked specifically about protocols for a complex cyber-biological attack, they referred us to the FBI. We reminded them that we are the FBI.

There appears to be no coordinated cyber-biology division or agency with jurisdiction over both cybersecurity and biosecurity. We cannot find any agency with protocols or plans to deal with a sophisticated attack that begins as malicious computer code and results in genetic code designed to act as a bioweapon. We appear to be in the early stages of a massive bioterror attack on American soil, and we have no central point of contact, no established protocols, and no strategy for how to contain this imminent threat.

Please advise.[1]

# PART FOUR The Way Forward

| FIFTEEN |

# A NEW BEGINNING

If you drive south from the FBI's San Francisco Field Office and connect in Half Moon Bay with Route 1, the road opens up to breathtaking sapphire blue whirlpools and jagged rocks jutting out from the ocean. There, the Pacific shoreline is edged with rough sand dunes, tall grasses, and forests of old-growth redwoods, cypresses, and pines. Near Monterey, the road forks, winding past yellow and orange wildflowers and then, finally, to the Asilomar Conference Center, a retreat nestled below the canopy of trees and built around the idea of integrating the natural environment with one designed by people.

Near the end of the nineteenth century, women were beginning to enter the workforce and taking low-paying jobs in factories and offices. Back then, three feminists headed the San Francisco chapter of the Young Women's Christian Association (YWCA): Ellen Browning Scripps (a publisher), Mary Sroufe Merrill (an activist, philanthropist, and writer), and Phoebe Apperson Hearst (a prominent suffragist and philanthropist, and mother of publishing magnate William Randolph Hearst). They had chosen that same stretch of shoreline for an outpost of the YWCA—but they had bigger ambitions. All were wealthy, so when it came time to build, they could have hired any of the top male architects of the time. Instead, they chose a little-known engineer and architect named Julia Morgan to design a small campus. Helen Salisbury, a Stanford student, won the contest to give the center its name, a hybrid of two Spanish words: *asilo*, meaning retreat or refuge, and *mar*, for the sea. When Asilomar held its first women's leadership conference in 1913, it was more than a "refuge-by-the-sea" and a simple outpost of the YWCA. Asilomar was a promise. A place where women would learn from each other and network with other progressive thinkers, who in time would grow to include a select group of men. Those who gathered at Asilomar sought to strip American society down to its most basic elements, and then reassemble it from the ground up to be inclusive, equal, and primed for a better future.[1]

Scripps, Merrill, and Hearst believed everyone had a sacred duty to question the powerful systems that governed their lives, even when that led to great uncertainty. They understood that, as people made advances in science and technology, life would need to be reimagined, again and again.

BY 1973, NOT far from Asilomar, research that would soon prove hugely influential was underway. Scientists at UC San Francisco and Stanford were experimenting with restriction enzymes to cut long strands of nucleotides into smaller bits of genetic letters and then insert them into other cells. They were hoping to create a process to swap DNA between different species. The resulting technology, called recombinant DNA, or rDNA, had profound implications. If scientists could exchange DNA in bacteria, which organisms might be next? One concern: microbes that cause cancer in mice could theoretically be transferred to, say, horses, and then, what if horse viruses made their way into humans? This unlocked a terrifying new possibility: researchers could, knowingly or unintentionally, create new diseases about which little was known, and for which there was no protection or cure. (Remember, in this era, there was no genetic sequencing machine, and cracking the code on a novel pathogen was a long and arduous process.) There was also no way to predict how these edited organisms might behave in the wild, or how they could evolve. But one thing was certain—humans had become something like gods. They had not just reimagined life. They had re-created and transformed it.

One of the researchers involved in this discovery, Stanford biochemist Paul Berg, sent a cautionary letter to the journal *Scientist* shortly after he first synthesized an rDNA molecule in 1972. "Several groups of scientists are now planning to use this technology to create recombinant DNA from a variety of other viral, animal, and bacterial sources," he wrote. "Although such experiments are likely to facilitate the solution of important theoretical and practical biological problems, they would also result in the creation of novel types of infectious DNA elements whose biological properties cannot be completely predicted in advance."[2] Berg had been meeting with other prominent scientists, including the biologists Maxine Singer, David Baltimore, Norton Zinder, and James Watson. Watson was then the director of the Cold Spring Harbor Laboratory, one of the world's leading biological research centers. They were worried about the potential dangers of rDNA molecules, which they knew could lead to self-replicating viruses, hazardous bacteria, or even bioweapons that could have catastrophic effects. But they also recognized rDNA's potential. If research continued, and if scientists learned how to harness the technology safely, it could unleash enormous powers to improve and extend life: to make synthetic insulin, create antibiotics, and invent new therapeutics that hadn't even been conceived of yet. Berg and his colleagues called for a moratorium on further experimentation until there was a set of principles to guide the use of gene editing research.[3]

That led them to two big questions. What should those principles be? And who would decide them? There were geopolitical issues to consider. US troops had just pulled out of Vietnam, and the Soviet Union was laying groundwork to install communist regimes in Southeast Asia, Latin America, and Afghanistan.[4,5] The United States and China had not yet established diplomatic relations.[6] If the group included only American scientists, other countries were likely to ignore or reject the resulting principles. There were moral, ethical, and religious considerations. Doctors in England were experimenting with a new procedure that would create an embryo in a "test tube."[7] This rattled theologians, who were woefully unprepared to address the profound moral questions the development raised.[8] A set of principles would further amplify long-standing religious beliefs about how life was formed—especially for those who contended that manipulating or destroying genetic material was implicitly a sin—which could hamper research rather than help it along. If the group consisted only of scientists, its principles could later be challenged by politicians, who could reasonably argue that any laws, including those regarding life, had to be determined by governments, not scientists.

Berg and his peers knew that a broad set of stakeholders would have to come to a consensus about the principles to reduce the risks inherent in this type of research. They decided to host a conference, on February 24, 1975, that would focus on two basic questions:

1. How should the protection of scientific freedom of inquiry be balanced with the protection of the public good?
2. How should decisions about scientific research and its technological applications in society be made, especially in climates of uncertainty?[9]

They came up with an international list of elite molecular biologists, journalists, doctors, lawyers, and other key professionals and invited them to Asilomar—that place of radical reimaginings—to create the path forward for engineered life.[10]

As Berg and Baltimore strode on stage at Asilomar for the opening session, they knew that not everyone in the room was familiar with what rDNA was. So they began by explaining the technology in clear terms, without exaggerating or sensationalizing its implications. But they also made the gravity of the event very clear. This group of individuals, who had come from the United States, the Soviet Union, West Germany, Canada, Japan, England, Israel, Switzerland, and elsewhere, were already being called biotechnology's version of a Constitutional Convention.[11] Accordingly, Baltimore closed the session with an ominous observation: if this group of people couldn't reach a consensus on how to use rDNA, no one else could.

The organizers also had secondary motives. This, and other emerging biotechnologies, would eventually attract attention from lawmakers. They and the general public would have a difficult time understanding rDNA.

Among those who didn't understand it, misinformation would spread quickly. Those organizers knew that science needed to be self-governing, but in order for that to happen, scientists and researchers needed to earn the public's trust and satisfy lawmakers' safety concerns. If this group at Asilomar, made up of people carefully drawn from many different disciplines, could openly debate their beliefs and come to a consensus, the scientists would have demonstrated their ability to balance scientific self-interest with voluntary self-restraint.

That's why more than a dozen journalists—from the *New York Times*, the *Wall Street Journal*, the Canadian Broadcasting Corporation, *Frankfurter Allgemeine*, and even *Rolling Stone*—had been invited, too.[12] Those journalists were expected to write about all of the deliberations, not just the final outcome of the meeting. Which meant that any bickering, name-calling, and nasty exchanges between scientists would be included in the published accounts that politicians and the public read. Scientists, who typically operated behind the closed doors of a lab, published difficult-to-read academic research, and generally avoided the spotlight, were concerned that a public debate would lead to new public scrutiny for biotechnology. Berg and his peers, though, anticipated a different outcome. If the public had a better understanding of what rDNA was, and if they knew scientists were working hard to prevent worst-case scenarios, then they would trust both the scientists and the science.

The organizers were right. The participants reached a consensus, agreeing on a set of restrictions and safety protocols to be put in place before rDNA research resumed. Official guidelines were issued shortly after. *Rolling Stone* published a sprawling, first-person account of Asilomar. Musician Stevie Wonder and geneticist James Watson were both featured in that same issue. A very 1970s psychedelic illustration of Wonder graced the cover: colorful abstract shapes were reflected in his sunglasses, and he was wearing headphones, a shaggy brown coat, beads, and a multicolored, poufy cap. Inside the issue there was a black-and-white photo of an awkward Watson in a rumpled sweater during a coffee break, listening to another participant (who was wearing an only slightly less rumpled sweater).[13] Most importantly, for the past forty years, rDNA technology has led to enormous scientific progress, with no negative public health repercussions and, most remarkably, without any epidemics of misinformation until very recently. Scientists earned the public's trust by proving that they could analyze risk, reach consensus, and self-govern their work. A new era for science, transparency, and public policy began at Asilomar.

Given He Jiankui's CRISPR experiments, misinformation about mRNA vaccines, and the prospect of developing human-animal chimeras, there are now calls to convene a new group of stakeholders at Asilomar to debate the risks and rewards inherent in synthetic biology.[14] But the world is a much different place now than it was in 1975. There are many biotechnologies that allow us the ability to fundamentally alter life, and advances in artificial intelligence, computer network infrastructure, and 5G and 6G wireless technology enable us to engage in new kinds of R&D. These are poised to boost innovation and a steady stream of new commercial products. It would be a mistake to hold a conference devoted to the implications of CRISPR without also debating the risks and rewards of AI's deep neural networks. Trying to achieve consensus across the matrix of technologies that comprise synthetic biology would be a serious challenge. Also, at this stage the patent landscape is confusing, and legal battles are still being waged in courts. Some scientists who would be invited to build consensus on the future of synthetic biology are currently suing each other.

While technology has advanced, so have the global ambitions—and complications—of the countries that lead its development. Russia is no longer a collaborator in biotech. China has prioritized synthetic biology in its quest for global science and tech hegemony. The US government hasn't been able to maintain a consistent stance on science and tech policy as administrations have moved in and out of the White House. There are also far more investors funding research now. Life sciences is one of the biggest and most attractive sectors for venture capitalists, hedge funds, and private equity. Any present-day Asilomar would need to include the heads of investment firms whose success depends on quickly bringing commercial products to market, and who are likely to harbor biases against long-term risk modeling.

When Asilomar was held in 1975, President Richard Nixon's speechwriters had just begun using the term "the media" in a derogatory way, in order to sow distrust in journalism.[15] Today, trust in media is at an all-time low, and the social media currencies of attention and influence reward consumers for posting salacious and sensationalized content.[16] If a new Asilomar were held today and proceedings were reported by journalists, what's the over-under on how quickly momentary exchanges would be taken out of context? The organizers of a new Asilomar would have to assume that any story about the proceedings, no matter how dutifully reported, would be twisted into mistruths online.

Meanwhile, as we write, three events shaping the future of life are underway. California lawmakers have proposed a new rule requiring mail-order DNA companies to perform biosecurity screenings; Ginkgo Bioworks went public with a $15 billion valuation; and anti-vaxxers had just taken over college campuses protesting new policies that required students to get COVID-19 vaccinations before returning to school in the fall of 2021.[17,18] The field of synthetic biology is evolving fast, but the foundations that support it—legal frameworks, the bioeconomy, and the public trust—are uncertain.

Absent a new Asilomar, some policy makers want detailed roadmaps defining the future of synthetic biology. Most often, these mirror economic roadmaps that imagine development, milestones, and measurable outcomes along a linear timeline. But science is nonlinear, especially when you're dealing with emerging technologies. Breakthroughs lead to advancements, but most of the time experiments fail. Discoveries usually only happen after many twists, turns, cul-de-sacs, and dead ends.

There is a way to set synthetic biology on a positive course for the future, one that owes much to the basic questions that Berg and his colleagues posed, and the example set by Scripps, Merrill, and Hearst. We cannot know exactly how the future of synthetic biology will unfold. But we can lean into uncertainty and learn from it if we ask questions that begin with "How should . . . ," "What if . . . ," and "Can we . . . ," and that end with ". . . for the public good." Doing so will likely require imagining yourself in a future that contradicts what you believe today. It will be uncomfortable. It will require courage. Widening your perspective in order to make informed decisions about the future, and then taking steps toward it no matter where your inquiry leads, are radical acts.

For synthetic biology to achieve its greatest potential while minimizing dangerous risks, we need to imagine ourselves in such an unfamiliar future, one in which our regulatory approaches, geopolitical agreements, and investment strategies look different than they do today. In that future, trust will stem from inclusivity, communication, and accountability. Scientific knowledge and understanding will be democratized, religion will coexist with science, and politics will clear a path for innovation. (Yes, we know this sounds like an impossible, fantastical plan to re-engineer society.)

This book is our version of Asilomar. We've invited a global group of stakeholders—you, and everyone else reading this book—to learn about synthetic biology technologies and the events leading up to this moment. You have met researchers, listened to their squabbles, and heard their perspectives. We've introduced you to some of the investors and companies within the bioeconomy. We have challenged your ideas as each chapter has unfolded. We've questioned your existing thoughts about how scientific research should be conducted and how we ought to make decisions about the future applications of synthetic biology technologies.

How should the protection of scientific freedom of inquiry be balanced with the protection of the public good? How should decisions about scientific research and its technological applications in society be made, especially in climates of uncertainty? What follows are our recommendations for global cooperation, regulation, business, and the synthetic biology community. These are intended as starting points. They are opportunities to continue asking questions and to build consensus.

## FINDING COMMON GROUND

Every time an exponential technology is introduced, a race ensues, especially if that development brings significant economic and national security implications. That was true for space exploration (with the United States and the Soviet Union vying for dominance), it's true for artificial intelligence (with the competition this time between the United States and China), and it is now true for synthetic biology. The winner reaps great benefits: the ability to direct capital investment, to attract top academic talent, to set the pace of innovation, and even to dictate global standards.

In Chapter 3, we gave a brief history of AI, which was conceptualized as early as the 1820s and given its name in 1956. The first wave of AI technologies came between 1960 and 1980, and this enabled the formation of a new business ecosystem, attracted talent and investment, and created much of the invisible infrastructure that now powers everyday life—the antilock brakes in your car, the fraud detection systems monitoring your credit cards, and so on. Yet AI is now moving along different developmental tracks, with decidedly different intentions. The United States has had no strategy or integrated policy on AI, and no plan to guide its objectives or growth. This left private sector companies, which often prioritize shareholders above the public interest, to make all the decisions. In practical terms, this resulted in sacrificing consumer privacy and selling data to unscrupulous third parties. It also resulted in crucial products and services, such as Facebook and YouTube, that are riddled with algorithmic bias.

Large companies have always lobbied to influence policy and regulation. But the tech giants have amassed once unimaginable power and wealth, and in doing so, they have made key decisions with significant diplomatic and geopolitical impacts. Some have even established their own corporate foreign policy departments. Microsoft's president, Brad Smith, regularly meets with heads of state and foreign ministers to discuss emerging cyberthreats, and to explore such issues as how to close the digital divide in developing economies. In 2017, he introduced a Digital Geneva Convention, an international treaty to protect citizens against state-sponsored cyberattacks.[19] Microsoft's Digital Diplomacy Group actively works on a tech-focused approach to foreign policy, with dozens of policy experts helping to develop international accords for cybersecurity and to shape local regulations, among other activities. It even hosts closed-door meetings with diplomats on human rights.[20]

The company understands that corporate foreign policy is good business—it builds trust and assists in long-term planning. Facebook, Apple, Google, and Amazon all employ similar strategies. Now, consider the long-term implications of tech companies influencing geoeconomics. What if a company like Facebook had priorities that differed from the priorities of its government? What if it promoted a policy during these diplomatic meetings that government policy makers had not yet addressed, or worse, that stood at odds with US policy? These scenarios, which look ever more plausible, have led to increasing challenges from state and federal lawmakers and investigations by regulatory agencies. They have drawn the ire of US presidential administrations. Now, those companies—America's most formidable AI players—are under assault for accumulating too much power and wealth. And the tech industry, investors, and government will spend the next decade arguing about it.

Meanwhile, in China, AI's development is led by its three major tech companies—Baidu, Alibaba, and Tencent, collectively referred to as the BAT—and its varied academic institutions. The BAT companies may be publicly traded, but they are guided by the Chinese Communist Party, whose views on privacy, surveillance, and human rights are drastically different from those of the United States and its allies. Beijing is positioning AI to perfect its authoritarian regime, both domestically and via policies such as the Belt and Road Initiative, which trades infrastructure development in emerging markets for debt. AI's superpowers, China and the United States, view AI as a linchpin to national security, economic growth, and military dominance. But the risks inherent in an unconstrained AI race are obvious, providing plenty of reasons why the United States and China should be motivated to forge a relationship where both can succeed.

The developmental track of synthetic biology parallels what we've seen in AI. In fact, some of the same players who built our modern AI economy are now deeply involved in building the bioeconomy. Microsoft is researching DNA storage and building automated technologies to support biotech foundries. In the past few years, Bill Gates has advocated for investment in synthetic biology to combat global hunger and climate change.[21] Jeff Bezos is backing several synthetic biology companies, and his space company, Blue Origin, would benefit from such tools and technologies, which can help humans to survive off-planet.[22] Former Google CEO Eric Schmidt invested $150 million in the Broad Institute to hasten the convergence of AI and biology.[23] Academics may conduct the research, but the commercial sector is providing the funds that will speed us toward the new innovations. Funds, of course, mean influence, especially over the direction of the research.

China has made it abundantly clear that it plans to achieve international supremacy in both synthetic biology and artificial intelligence. Its state policies call for the country to be the "world powerhouse of scientific and technological innovation" by 2050.[24] The Chinese Communist Party has been working tirelessly to erode the United States' long-standing technology advantage for the past decade.[25] The Chinese government launched its National GeneBank, which aims to become the world's largest repository of genetic data, in 2016.[26] The CCP eyes a strategic value in DNA—for drug discovery, to advance its agricultural sector, and to maintain social order—and its efforts are aided by BGI, a leader in cheap gene sequencing we've mentioned elsewhere in this book.

There appear to be links between BGI's research and China's military—the People's Liberation Army (PLA), which maintains supercomputers to process genetic information. It's well documented that the PLA sponsors research on gene editing and performance enhancement, among other offensive capabilities, and PLA leaders have specifically cited synthetic biology as a future domain for warfare. Some have spoken publicly about engineering brain control weapons. There are a strikingly large number of CRISPR trials taking place at medical institutions affiliated with the PLA.[27,28]

We want to be clear: many scientists working in China do not share the ambitions of the CCP and the PLA. Globally, the synthetic biology community is open and collaborative, and that includes the generosity of too many Chinese scientists to list here. Yuan Longping, an agricultural scientist, developed rice hybrids in the 1970s that helped alleviate famine in parts of Asia and Africa.[29] Rather than confining himself to a lab, or joining the CCP and taking a plum government position, he dedicated his life to the eradication of hunger, spending time in the fields talking to farmers. He mentored the next generation of scientists all over the world.[30] You may recall Dr. Li Wenliang, an ophthalmologist working in Wuhan, tried to warn his colleagues about SARS-CoV-2 early on, using Weibo, China's carefully monitored social media platform.[31] As his posts went viral, he acknowledged that he would face severe punishment by the government, yet he continued to post—eventually from his hospital bed—right up until he died. Zhang Yongzhen and his team sequenced the SARS-CoV-2 genome and went to great lengths to ensure that the greater biology community would see it—even posting in a public forum.

The scientific community may be open and globally connected, but the CCP wants to repatriate Chinese talent. As of this writing, more than 250,000 life sciences professionals have answered that call.[32] China is now among the world's leaders in patents and academic publications. It has also made a giant leap in high-tech manufacturing, thanks to a "Made in China" industrial strategy that depends in part on biotechnology advancements.[33] The government is actively building bio-enterprise capacity, education, and life sciences parks around the country, even as its intellectual property laws and regulatory environment still do not measure up to international standards. While the world was shocked to learn that He Jiankui's CRISPR-edited embryos resulted in live births, it is likely that the CCP was aware of his research. He wasn't exactly working in secret, and China has the most sophisticated surveillance systems in the world. The environment in China allowed, and possibly even encouraged, the sort of genetic engineering experiments that would have been unthinkable anywhere else.

It is abundantly clear that China isn't interested in becoming the world's biotech factory. Rather, it intends to be the dominant global superpower in both synthetic biology and AI. By 2030, China is expected to be the world's largest economy measured by GDP. By 2050, it could also be one of the largest holders of patents and intellectual property, and also the first country where all new babies are sequenced at birth. It has substantial incentives to make sure the bioeconomy is built on its own terms. China's population is enormous, and the world is facing climate-related migration and food production challenges. If China succeeds, it will become the world's top exporter of sequencers, pharmaceuticals, agricultural staples, and solutions to mitigate pollution and extreme weather events.[34]

Other important developments are underway in countries where laws governing genetic engineering, biotechnology, and personal data vary wildly from international norms. India will be the world's most populous country by 2050, and it could be among the world's largest economies. It will also be a major food producer. India's sheer market size, scale, and importance as a global food producer means that its influence on the developmental track of synthetic biology is inevitable. India's government established a Department of Biotechnology in the 1980s in order to develop strategies for the future of genetic modification and other technologies.[35] But the country's notorious bureaucracy hampered the department's efforts to develop and enforce regulatory frameworks. At the same time, pharmaceutical manufacturing plants were caught cutting corners on production and falsifying data to meet profit targets.[36] Currently, India is home to many talented scientists, technologists, and entrepreneurs, but it lacks both a national strategy and global trust in its ability to develop and manufacture high-quality biotech. Its lax oversight could threaten everyone. It could be motivated to enact a wide array of regulations in order to attract investment and find markets for its products—and then, given its history, fail to enforce them.

Israel and Singapore are both building biotech capacity, collaborating with other countries, and seeking foreign investment. Both have adopted policy approaches to spur innovation. Israel launched an "Innovation Box" program, which seeks to persuade multinational corporations to move (or at least colocate) their R&D operations to that country, thanks to corporate tax breaks and other financial incentives.[37] Its Tzatam program provides equipment and other assistance needed for R&D work in synthetic biology.[38] Germline gene editing in humans is prohibited, but research on plants and animals is encouraged, and stringent risk-assessment processes govern which products can be sold commercially. Singapore created state-of-the-art policies to drive biotech innovation, and those policies are integrated with its education, economic, health, and agricultural sectors. It's no surprise that the world's first biofoundry-grown meats are now on the market in Singapore.[39]

And the European Union? It adopted strict regulations on genetically modified foods in 1997, and public trust of synthetic biology technologies among Europeans is low. A 2020 Eurobarometer study found that two-thirds of Europeans won't buy genetically modified fruits, even if they taste better or were grown more sustainably than non-GMOs. In 2018, France strengthened its regulations regarding CRISPR technology, making it subject to the same regulations as GMOs.[40] However, older techniques, such as exposing plants to radiation to trigger random mutations, aren't covered. This has had a chilling effect on the scientific research community in Europe and the United Kingdom. Cross-border research projects that used CRISPR to edit plants were immediately shut down. One scientist, who used CRISPR to make camelina oilseed plants to produce more healthful omega-3 oils, was told that the regulatory status of his field trial had changed—while his plants were still in the ground.

It's clear that science and science policy are out of sync. The planet-scale challenges that synthetic biology can help solve—our climate emergency, plummeting biodiversity, food scarcity, and the emergence of new pathogens—demand global collaboration. Yet countries are incentivized to compete for market share, and potentially even to develop the kinds of bioweapons that are not yet governed under international treaties. There is no way to prevent what nature will do on its own. Nor is there any way to predict all the ways in which humans will develop dual uses for technology. But we can offer three global recommendations to reduce the risks associated with synthetic biology.

### RECOMMENDATION #1: BAN GAIN-OF-FUNCTION RESEARCH

When a technology gets invented, people create off-label uses for it. We should assume that the same will be true for synthetic biology. This is why gain-of-function research needs to be banned. Recall that gain-of-function research makes viruses more dangerous (Chapter 7). Let's just call it out for what it is: bioweapons development.

Even if every country, laboratory, and DIY biologist on Earth agreed to stop using synthetic biology technologies, nature would still invent its own dual-use issues. Consider the bacterium *Yersinia pestis*, which began attacking the army of a Mongolian king, Jani Beg, in the 1340s. Jani Beg's soldiers might have been winning wars against their western foes, but they were losing an internal immunological battle to this deadly pathogen. Outbreaks spread to troops in Constantinople, then to Sicily, and eventually to Marseille. By the time it reached the Persian Empire, *Y. pestis* was known simply as the Black Death. The bacterium evolved over hundreds of years, made its way into fleas, soil, mammals, and, eventually, Europe's population. It caused a third of the people in Europe to die horrible, gruesome deaths.[41] But there are plenty of other examples: Malaria. Rabies. Tuberculosis. Ebola. Or COVID-19—if you believe its origin was indeed natural. There is absolutely no reason for us to give nature a helping hand.

Given the modeling and sequencing technology readily available today, there is little need to conduct gain-of-function research to prepare for viral outbreaks. When Ron Fouchier "mutated the hell" out of the H5N1 bird flu microbe in 2012, it was to build virus models for research purposes. At the time, some scientists feared that if a novel pathogen was discovered, it would take too long to sequence the genome. If those in the scientific community had genomes for hyper-transmissible, deadly versions of the virus in advance of an outbreak, they believed they'd be better positioned to develop vaccines and medical treatments quickly. But the vast majority of scientists and others working in the synthetic biology community were alarmed when they learned about Fouchier's work. With the tools of synthetic biology constantly improving, this type of research is even more dangerous—and, we believe, unnecessary.

A decade after Fouchier's experiment, our computer systems are exponentially more powerful and our genetic databases are enormous. Sequencers are capable of revealing genetic code in a matter of hours. Analysis and models of probable mutations can all be done in computer simulation. Meanwhile, even the most secure biolabs, in the United States and elsewhere, have been cited for safety issues, ranging from poor inventory management to insufficient decontamination of wastewater. That we cannot yet easily rule out COVID-19 being the product of gain-of-function research in Wuhan (as of this writing in mid-2021) screams that the risks far outweigh the value this research has for public safety. Moreover, COVID-19 revealed how grossly unprepared we were to manage a moderately infectious, moderately deadly virus. Consider what might have happened if the virus had been just slightly more infectious and more deadly.

In December 2017, the Trump administration released new guidelines clearing the way for government-funded gain-of-function projects intended not just to monitor for new potential pathogens, but to encourage the study of intentional gain-of-function mutations. To other nations, this broadcasts a clear message: the United States is working on viral bioweapons. The last thing we need right now is a biological arms race. It's worth noting that the companies that make vaccines haven't publicly called for gain-of-function research or indicated that the research would assist them in ramping up supply chains for future vaccines.[42]

Banning gain-of-function research isn't tantamount to stopping work on synthetic viruses, vaccines, antivirals, or virus tests altogether. We are surrounded by viruses. They're important and integral to our ecosystems. They can be harnessed for beneficial functions, which include precision antibiotics for hard-to-kill microbes, cancer treatments, or delivery vehicles for gene therapies. But we should monitor this type of work as closely as we monitor the development of nuclear technologies.

### RECOMMENDATION #2: CREATE BIOTECH'S BRETTON WOODS

Countries typically come together during a crisis, not before one. It's easy to agree on danger. It's far harder to agree on a shared vision and a grand transformation. But countries could be incentivized to collaborate for public good, because they have an overwhelming interest in, say, developing their bioeconomies, instead of spending resources to create new tools for biowarfare.

One model is the Bretton Woods agreement, a 1944 pact between the Allied nations of World War II that laid the foundation for a new global monetary system. Among the agreement's provisions were plans to create two new organizations tasked with monitoring the new system and promoting economic growth: the World Bank and the International Monetary Fund (IMF). The Bretton Woods nations agreed to collaborate: if one country's currency became too weak, the other countries would step in to help; if it was devalued beyond a certain point, the IMF would bail that country out. They also agreed to avoid trade wars. But the IMF wouldn't function like a global central bank. Instead it would operate as a sort of free library, from which its members could borrow when needed, while also being required to contribute to a pool of gold and currency to keep the system running. Eventually, the Bretton Woods system included forty-four countries, which came to consensus on regulating and promoting international trade. The collaborative approach worked well, because all members stood to gain or lose if they violated the compact. The Bretton Woods system was dissolved in the 1970s, but the IMF and the World Bank still provide a strong foundation for international currency exchange.[43,44]

Instead of monitoring and regulating a global pool of money, the system we propose would govern the global pool of genetic data. Member nations would agree to use a blockchain-based, immutable ledger and tracking system to record genetic sequences as well as standardized parts, orders, and products. Whether scientists were de-extincting the Tasmanian tiger, using CRISPR to enhance collagen production in adults, or discovering a new pathogen, the genetic information used or created would be entered into this shared global system. Facility and product inspections, which would have to meet strictly enforced standards, would also be required and entered into the system, thus creating a chain of accountability. For instance, reputable companies follow strict biosecurity precautions. Every month, Twist Biosciences screens thousands of genetic sequence orders from academic labs, pharmaceutical companies, and chemical manufacturers, looking for anomalies. Every now and then it uncovers a dangerous order (which, it turns out, typically stems from a customer's inadvertent error). But we wrote our FBI scenario because this is not true of all of Twist's peer companies. This kind of global system would require companies to screen synthetic gene orders against various DNA databases housing sequences of regulated pathogens and known toxins, and then authenticate buyers and record transactions in a public database.

The global pool of genetic data includes DNA, which reveals our most sensitive and personal secrets. Insurance companies, the police, and adversaries would be intensely interested in that information. At least seventy countries now maintain national DNA registries, some of which include data that were collected without gaining informed consent. The current approach to national registries positions DNA as a policing tool while missing the opportunity to pool genetic data for globally scaled research projects that could benefit us all.[45]

A tiny country of just 1.3 million people demonstrates a better way forward.[46,47] From a fragile perch in northern Europe, uncomfortably close to a hostile Russia, Estonia built what has long been considered one of the world's most advanced digital ecosystems. Its state-issued digital identity allows residents to safely handle online transactions with government authorities, tax and registration offices, and many other public and private services. Citizens have voted electronically since 2005, using their digital ID for authentication. That same digital ID serves as a backbone for Estonia's health system, which connects citizens and their centrally stored personal health and medical records to doctors and health-care providers. Estonia's digital ecosystem also makes it easier to do data-intensive genetic research. The country's BioBank includes genetic and health information for 20 percent of its adults, who consented to opt in to genetic research programs. Estonia's system offers them free genotyping and related education classes, which—bless the Estonian ethos—people actually attend. That digital ID system also guarantees participants security and anonymity.[48]

In a biotech Bretton Woods system, member countries could build a similar blockchain-based digital ID system to create an unchangeable ledger of personal genomic data for research programs. Estonia's model for informed consent is a good model for member nations of our proposed system. Member nations would then contribute a percentage of their population's genetic data into a global pool. Such a system would encourage responsible use and development of genetic data and incentivize accountability. A standard system for genetic-sequence storage and retrieval would make audits easier and more scalable.

## RECOMMENDATION #3: REQUIRE A LICENSE

The modern car is a powerful technology. Every country requires training, driver's licenses, approved safety measures, registrations, monitoring, and regulatory enforcement for both drivers and automotive manufacturers. Each one maintains a national registry of licensed drivers that is regularly updated. More specific certifications and licenses are required for motorcycles, large trucks, and delivery vehicles, and we expect autonomous vehicles to have their own special versions in the near future as well. Drivers are expected to pass written and driving exams to prove they know rules of the road. Currently, 150 countries require foreigners who want to drive a vehicle during their stay to first get an International Driving Permit, which requires an application form and a valid license from their home country.[49] Manufacturers, meanwhile, must pass dozens of different inspections to sell their vehicles: airbags must deploy correctly, brakes can't seize up, seat belts cannot fail. Vehicles are tested using computer simulations, then in closed-environment simulations using crash-test dummies, then on closed outdoor courses using humans. The vehicles themselves are required to display license plates and environmental inspection certificates. Once those cars are on the road, radar monitors for speed, red light cameras make sure people stop when and where they should, and local law enforcement officers patrol for reckless, drunk, or just really bad drivers. If you try to sell a used car to someone, the paperwork and data chain of inspections, licensing, and registration begin all over again.

Couldn't we build a similar licensing system for synthetic biology, one that covered everyone from DIY biohackers to professional researchers, and that governed products and processes, required rigorous testing, and ensured that trade and commerce were closely monitored? It's a reasonable approach. And it isn't our idea—it's George Church's. In his book *Regenesis*, Church recommends a "suite of safety measures comparable to those that we now have for cars."[50]

We'd take that a few steps further. An international licensing system would confer a certification, and would therefore attract more people to the field, which today competes to some degree with AI. Since biology isn't static, the certification would need to be maintained through continuing education, thereby ensuring that hobbyists would have to prove they knew the latest information. Countries could subsidize the costs of licensing, or even build policy into their education programs, to attract more young people into the bioeconomy. Licenses would be internationally recognized by member countries, which could help to facilitate collaboration between researchers. An international licensing system would also cover safety standards for manufacturers of synthetic biology equipment, biofoundries, and commercial enterprises working in all areas of the discipline. A future sequencer or synthesizer could have screening systems built directly into the hardware, which could theoretically make it more difficult to intentionally or accidentally design a destructive organism. A future safety measure might include an emergency brake: an auto-destruct feature coded into cells should they migrate outside their lab environment. A licensing system would also promote standardized systems and interoperability, which would help bioeconomies flourish in every country.

One thing is certain: the path we are currently following promotes geopolitical tensions, unhinged competition, and conflicting regulations. It will lead to global conflicts. Our version of Church's alternative promises both safety and economic benefits through cooperation.

## THE UNITED STATES NEEDS BETTER SCIENCE AND TECHNOLOGY POLICY

The United States may now be a global leader in synthetic biology, but we've created tensions between researchers, investors, and local regulators. For starters, our regulatory frameworks manage to hamper innovation while also failing to protect citizens against future harm. The Coordinated Framework for the Regulation of Biotechnology stipulates that three agencies—the EPA, the FDA, and the USDA—each play a role in regulating biotechnology, but the system isn't updated regularly. This leaves gaping vulnerabilities in our oversight. We developed our FBI Field Office scenario after researching which government agency was in charge of monitoring both synthetic biology products and potential hacks into genetic data.[51] We were curious because, in November 2020, academics at Israel's Ben-Gurion University of the Negev created a similar novel cyber-biological attack that could have tricked a scientist into creating a dangerous toxin-producing genetic sequence.[52,53] Understandably, we were alarmed about this, so we spent three full days reading policies at the US Department of Homeland Security and poring over National Security Council documents to determine who would have jurisdiction over biological malware. We interviewed key sources at the Department of Defense, the State Department, the Government Accountably Office, and the CDC, as well as national security analysts and congressional staffers. We were referred to different people, and various agencies, until one high-level staffer finally told us the answer: the United States is totally unprepared for a cyber-biological attack.

The Trump administration relaxed the Coordinated Framework in 2019. It now allows noncommercial experimentation without much scrutiny instead of providing specific guidance for the evolving array of synthetic biology technologies. Without intervention, we're likely to merely tack on more amendments and additional clauses to the Coordinated Framework, which will only cause more confusion and legal battles down the road. This situation reminds us of how the foundations of the internet were pieced together over time, leading to the systems we have today. The Internet Protocol is essential, as foundational as DNA. But the lack of centralized planning and coordination resulted in systems that have vulnerabilities, monopolies with control over the levers, and business models that incentivize profit over people. Synthetic biology shouldn't have to go through this organic learning curve.

The conditions that led to the mushroom problem that we outlined in Chapter 7 shouldn't exist in the future, and yet there is little effort being made to make sure that it doesn't. We need to create a system that promotes the responsible development of synthetic biology technologies and the bioeconomy. A bipartisan plan to establish a department dedicated to cyberbiosecurity and develop modern regulatory policy would be a start. That plan could ensure the safety of the biotech ecosystem and the viability of our long-term R&D funding targets, articulate the US vision for how synthetic biology technologies can spur economic development, and work to prepare our future workforce, augment our national security, and promote civic well-being. The key here is *long term*. Any plan must be able to withstand the cyclic changes in which party controls Congress (potentially every two years) and the presidency (potentially every four or eight years).

When complicated social issues arise, the federal government tends to let states do what they want. It's a way to preserve our democratic governing structure and prevent ideological hegemony. It's also a convenient way to pass the buck on a problem nobody wants to touch. In the early days of COVID-19, the federal government didn't purchase and distribute scarce PPE (personal protective equipment) supplies and ventilators, which led to a wasteful and bitter bidding war between states. Masks quickly became a political statement during the pandemic, with sharp divides that resulted in widespread protests. Both the Trump and Biden administrations refused to enact a national mask mandate, which left the job to state governors. Some governors, fearing voter backlash, left the decision up to city mayors and town councils. Masks became politicized, but the science is very simple: viruses spread through respiratory droplets when people sneeze, talk, cough, and breathe. Cover the source of the droplets, and it's harder to spread and inhale the virus.

What happens when the science is much more complicated? Key West has a horrible mosquito problem, which is causing the spread of the Zika virus. Gene-edited mosquitoes offered a potential solution. A group of researchers proposed editing male mosquitoes at the germline, which would prevent female offspring (which bite) from surviving. Germline editing bugs required the EPA to issue a permit, which it did. It then left community management up to the Islamorada Village Council, which, at the time, was made up of five people: a photographer, a retired real estate lawyer and part-time commercial fisherman, a local businessman, another local businessman, and a retired FedEx pilot.[54] These guys weren't scientists. Nonetheless, they were expected to hold hearings and make decisions about complex gene editing matters. The stakes were incredibly high. The tiny community could revolt against them. The pilot program could go horribly wrong. There could be unintended consequences for Key West's environment. It put those on the council in a terrible position. Leadership and a consistent, long-term national vision for the future of synthetic biology would reduce confusion in communities, enable local officials to make sound decisions, and create more opportunities for responsible bioeconomy growth.

## BUSINESSES: PREPARE FOR DISRUPTION

Synthetic biology will eventually intersect with every single industry sector, and therefore every business. Its advancements will change industrial materials, coatings, recycling, packaging, food, beverage, beauty, pharmaceuticals, health care, energy, transportation, and the supply chain. Synthetic biology will also change design (what and how we create), work (fewer sick days for employees), law (what and who we protect), news and entertainment (the kinds of stories we tell), education (what we teach), and religion (what we believe). Eventually, entire value chains will be transformed. Consider the current value chain for meat, which is long and costly. It involves breeding animals, providing them with feed, sheltering them, slaughtering them, processing the carcasses into different products, and preparing them for distribution. In the near future, cultured meat will drastically compress that chain into just a few links: sampling and storing tissue, cultivating cells, growing and texturing the meat, then readying it for distribution. All of this could soon be done within just one facility. This development will have wide-ranging impacts on refrigerated trucking companies and cold-storage warehouses, on the manufacturing companies that make the materials used to package meats, and on the tens of thousands of people who work in slaughterhouses. We're already seeing a version of this scenario happening in vertical farming, with companies such as Bowery Farming, Plenty, and Aerofarms bringing their computer-controlled indoor farms into urban centers.

In our experience, though, few companies are willing to develop their vision and strategy for technologies that might still be five to ten years away from widespread adoption. The longer companies wait to develop scenarios describing possible trajectories, the bigger risks they create and the more vulnerable they become to disruptors. Synthetic biology, like all other transformative technologies, will go through varying waves of innovation, failure, and success. Companies should still build capacity now to assess infrastructure, processes, and workforce skills. They'll need to evaluate their business models in order to determine how they might need to evolve. One question executives often ask is, "When, exactly, will synthetic biology disrupt our business and our industry?" Our answer: "when" is inconsequential. Companies need to identify inflections before they happen, and they must position themselves accordingly.

Companies operating within the bioeconomy must also remember that synthetic biology's ultimate stakeholder is our planetary ecosystem, and all of the living organisms within it. Academics have a tradition of peer review. Companies do not. To riff on the well-known saying, companies innovate fast, break rules, and make apologies later. Basic research is often incongruous with investor expectations. Investors, boards of directors, and marketers should give researchers the space and time to conduct research, studies, and field trials without making wild and premature proclamations about potential outcomes, and without rushing products to market. The success (or failure) of a business within the synthetic biology ecosystem impacts all companies. We spoke with John Cumbers, founder of SynBioBeta, about Ginkgo Bioworks going public just a few days after the deal was announced. He sounded excited. He also sounded concerned. "It's a big win, for sure," he said. "We all have this vision of a new platform that can revolutionize manufacturing industries, but the size of the valuation caught me off guard. It's a big bet on the future—a big bet in general—that impacts everyone in the field."[55]

Biotech companies also need to develop data governance policies that can be understood clearly by average citizens. In 2018, 23andMe announced a partnership with GlaxoSmithKline (GSK) that had been long in the making. GSK purchased a $300 million stake in 23andMe that allowed the pharmaceutical giant to use the startup's trove of genetic data to develop new drugs. The companies called it a "collaboration," but the deal didn't make it easy for millions of 23andMe users to opt out of a pharmaceutical giant's drug discovery process.[56] Unsurprisingly, consumers were angry: they hadn't agreed to volunteer for medical research that could earn a pharmaceutical company huge revenues. Even after considerable public outcry, most direct-to-consumer genetic analysis companies today sell consumer data to third parties. It's baked into their business models, but they obscure this crucial detail in confusing fine print. Some recipients of the data are major retailers, which now use DNA data to generate sales leads for online grocery shopping.[57,58]

What if a company that collects and houses consumer genetic samples is sold? (This has already happened with companies in the ancestry, cord blood, and fertility industries.) Sales of private companies are not uncommon, and in other industries buyers can include a menagerie of wealthy individuals, private equity, and trusts as well as other companies. What should happen to consumer data in the event of an acquisition? What happens when such data is bought and sold, and what if it is eventually sold to a foreign government? Data governance policies should be clear and understandable. Every effort must be made to earn and maintain consumer trust.

## COMMUNICATING SCIENCE MORE CLEARLY

Public trust is built on clear communication. In 2007, researchers at Carnegie Mellon, Stanford, and MIT conducted a joint study to learn what parts of the brain fire up when people go shopping. Researchers gave adults twenty dollars to spend at a specially designed online store while they were hooked up to a functional magnetic resonance imaging (fMRI) scanner. By the end of the project, the system was able to determine what parts of the brain were active when people were considering whether to buy a product; it could even predict whether they would ultimately choose to purchase it. The researchers published a sober-minded paper discussing their findings in the journal *Neuron*. It wasn't the sort of thing you would read before bedtime, unless you wanted to fall asleep quickly. They began by describing microeconomic theory (which "maintains that purchases are driven by a combination of consumer preference and price"). It was nuanced and complicated, however, and represented a significant achievement.[59] The universities behind it sought to win attention for their work, and the press office at Carnegie Mellon sent out an announcement that took a decidedly different tack. Its headline read, "Researchers Use Brain Scans to Predict When People Will Buy Products."[60] Eventually the study made its way to MTV, which declared "What's Better, Sex or Shopping?" MTV went on to explain that your brain thinks shopping is as tantalizing as a wild night in the sack.[61] Yet another example of how the people doing the science and those publicizing it are rarely in sync.

Scientists should assume that their studies and papers are being read by people who don't have the background to understand the details and context of their work. Scientific studies now get passed along from academic preprint servers (online repositories of papers that aren't yet peer-reviewed by the traditional academic journals) to peer-reviewed journals, then to press offices, journalists, regulators, adversaries (rival scientists and companies, countries, and individual bad actors), activists, investors, and, of course, people skimming headlines on social media. As the field of synthetic biology develops, scientists need to be able to communicate their work clearly. This means not ignoring the press office when they call them, and it means insisting that corrections be made when claims about their research are exaggerated, unclear, or just plain wrong. One solution is to find a way to get ahead of potential misinterpretations. Just before a team of researchers published a study that linked genetics to educational success on biorxiv.org, the preprint server for biology, it posted an enormous FAQ online in plain, clear language. It answered just about any question you could imagine; in fact, the FAQ is significantly longer than the paper itself.[62] More importantly, it's a heck of a lot easier to read. This could become a standard practice for researchers, preprint servers, and peer-reviewed journals alike: every paper could stand to include a one-paragraph summary written for non-researchers when published online (in addition to the standard abstract designed for researchers in the field). Coupling that with an FAQ could go far to prevent misinterpretations.

## SCIENCE HAS A RACISM PROBLEM

What if you agreed to participate in a scientific study, in which you would donate some of your blood to help identify a genetic link to a disease devastating the people in your community? You would expect your genetic samples to be kept private. Now imagine that you discovered, many years later, not only that you had been misled, but that other researchers were mining your DNA in ways that had never been disclosed to you. The Havasupai Tribe, which has lived for centuries in what is now Arizona, lived through this repugnant scenario.[63] Late in the twentieth century, the Havasupai were grappling with an increase in diabetes. They allowed researchers from Arizona State University (ASU) to conduct a study in 1990, hoping it would help them eradicate the disease from the tribe. The researchers gathered blood samples. But then, unbeknownst to the Havasupai, they changed the scope of the project to encompass genetic markers for alcoholism and various mental disorders. The ASU researchers went on to publish many papers in academic journals highlighting their results, and these articles resulted in news stories about inbreeding and schizophrenia among tribe members. The Havasupai were, understandably, horrified and humiliated, and they filed their first lawsuit against ASU in 2004. ASU paid for a private investigation and eventually settled the suit in 2010, returning the blood samples to the tribe and promising not to publish any more research.[64] But the experience deeply angered the Havasupai and other Indigenous peoples. The Navajo Nation, the second-largest group of Indigenous people in the United States, banned all genetic sequencing, analysis, and related research on its members. Their objections were absolutely warranted. But now there is a different problem: the pool of genetic data in the United States doesn't include Indigenous people.[65]

Our genetic data repositories are bereft of Black DNA, too. That's somewhat surprising, given that the first human cell line used in research was from a Black woman. In 1951, scientists retrieved cancer cells from a patient named Henrietta Lacks while she was being treated at Johns Hopkins Hospital in Baltimore. She had a large, malignant tumor on her cervix, which at the time was treated using radium. Most patients who underwent this treatment saw their cells die off, but Lacks's cells lived, doubling every twenty to twenty-four hours. Hopkins researchers decided to continue using her cells—which became known as the HeLa cell line—to develop various cancer treatments. But they never told Lacks or her family members about this decision. Nor did they compensate the family for her enormous contribution to the development of cancer therapies. (Finally, in late 2020, the family received a large gift from a nonprofit medical research organization.)[66]

Lacks's story is not the only one in which Black people were used in medical research without their knowledge. In 1932, the US Public Health Service launched a syphilis study at the Tuskegee Institute, a historically Black college in Alabama, which initially involved 600 Black men. In this study, 399 of the participating men had syphilis and 201 did not.[67] Researchers told all of them that they were being treated for "bad blood," a local term used to describe several ailments, including syphilis, anemia, and fatigue. In exchange for taking part in the study, the men received free medical exams, free meals, and, in a rather grim touch, burial insurance. By 1943, penicillin was used to treat syphilis, and it was widely available. The participants weren't offered any. At least 28 of the men in the study died. Hundreds of others suffered needlessly from painful sores and rashes, weight loss, fatigue, and organ damage. Taken together, the Henrietta Lacks saga and the Tuskegee syphilis study, along with many other cases that are less well known, make it easy to understand why some Black people are hesitant to visit doctors or participate in medical research.

As a result, the pool of genetic data in the United States consists mostly of people with European ancestry. A similar problem exists in the United Kingdom. A 2019 study conducted by MIT's Broad Institute, Harvard, and Massachusetts General Hospital reviewed the UK BioBank's genetic data to develop prediction scores for height, body mass index, type 2 diabetes, and other traits and diseases. These scores were meant to create baselines for doctors to use when treating patients and to help pharmaceutical companies develop new medicines. A dangerous pattern soon emerged: scores for people of European ancestry were 4.5 times more accurate than those for people of African ancestry. In the world's two largest English-speaking countries, we have a shocking dearth of understanding of health and disease in Black individuals.[68]

Genetic data are increasingly being used to research health. Without equity in our genetic databases, we will perpetuate gross inequalities in knowledge and care. To be fair, some steps are being taken to improve diversity in research. An Obama-era initiative called the All of Us Research Program began national enrollment in 2018 with a goal of collecting samples from a million (or more) Americans. As of December 2020, 270,000 people had contributed biosamples, and more than 80 percent of them came from communities that are historically underrepresented in biomedical research.[69] But there's still so much catching up to do.

The broader scientific community, which includes universities, publishers, and policy makers, has a diversity, equity, and inclusion problem, too. The organizations that support the biosciences, including American Association for the Advancement of Science (AAAS), the United Kingdom's Royal Society, and the nonprofit Open Access publisher PLOS, are overwhelmingly homogeneous. Nearly 80 percent of the AAAS leadership are white; 90 percent of the members of the Royal Society's editorial boards are white; and 74 percent of the editors employed by PLOS are white.[70] Editorial boards for the most prestigious peer-reviewed bioscience journals lack diversity in general: people of Middle Eastern or North African descent and Latinos are grossly underrepresented. The masthead of the peer-reviewed journal *Cell*, a destination for academic papers within synthetic biology, includes 15 editors, 7 staff members, and 119 advisory board members. Only one of them is Black.[71] Getting research published in top journals is academic currency, and that process typically requires personal connections between researchers and the journal editors who review manuscripts. If we're going to create a future of synthetic biology that represents all of us, we need to diversify the ecosystem.

## YOUR LIFE, REIMAGINED

The genesis machine is powered on and running. It is leading us inexorably toward a great transformation of our societies and our species. In the coming years, new genetic technologies will challenge your core beliefs. You will have to make choices about whether to sequence your DNA, and whether to vaccinate your child using messenger RNA. You will hear a lot about whether genetic selection—and enhancement—should be allowed, and who should have access to the technologies promising to improve life. You'll have to decide what you, too, think about these issues, and when the time comes, whether to enhance yourself. Climate change will impact your quality of life in ways that could change your job, your living conditions, and your community. When the time comes, will you switch to eating cultured meats? Do you trust genetically modified crops—and, if not, what would it take for you to change your mind?

Again and again, along with the rest of society, you will find yourself thinking about one of humanity's most difficult and ageless questions: What is life? As you search for answers, you'll have to take many developments in synthetic biology under consideration: synthetic insulin, Venter's minimum viable genome, a mostly woolly mammoth, a cluster of human pancreatic cells growing in a monkey, and, just possibly, one of your own skin cells, waiting to be converted into an embryo.

A global conversation is now underway, one that will shape the trajectory of synthetic biology in every community. You are now part of that dialogue. You are part of the genesis machine, and of humanity's great reimagining.

# EPILOGUE

BIOTECHNOLOGY GAVE US OUR FAMILIES.

We used our considerable experience researching and working within the field of synthetic biology to solve our fertility problems. We consulted with our colleagues, and with experts within their networks, and we employed the most cutting-edge technologies available. Andrew and his wife created two healthy babies—a girl, Rosalind, and a boy, Darwin—using IVF and, with Darwin, pre-implantation genetic testing. Amy and her husband created their daughter using a combination of genetic testing, ovulation induction agents, and acupuncture.

We understand how fortunate we are. We hope that ours is the last generation to endure these struggles, and that in the future technology-assisted pregnancy—genetic screening, sequencing, embryo selection, the many different gestation options open to every person—will be socially acceptable and widely available, rather than the desperate last option for people of means who have been unable to conceive.

A great transformation of life is underway. The genesis machine will soon determine how we conceive, how we define family, how we treat disease, where and how we make our homes, and how we nourish ourselves. It will help us fight our climate emergency, and it will enable biodiversity to flourish. It could create a better world for our children and perhaps help them explore other worlds, too. For their sake, we remain hopeful that synthetic biology will achieve its best possible futures.

# ACKNOWLEDGMENTS

MUCH LIKE SYNTHETIC BIOLOGY ITSELF, THIS BOOK MORPHED AND evolved as we explored the various dimensions and futures of biotechnology over the past decade. We have many people to thank.

**Amy**—*The Genesis Machine* is the result of hundreds of meetings, phone conversations, interviews, email exchanges, and wide-ranging discussions over wonderful meals. I am grateful to Arfiya Eri, Jake Sotiriadis, Jodi Halpern, John Cumbers, Kara Snesko, Frances Colon, Noriyuki Shikata, John Noonan, Masao Takahashi, Kathryn Kelly, Craig Beauchamp, Jim Baker, Bill McBain, Sewell Chan, Ros Deegan, Alfonso Wenker, Julia Mossbridge, Camille Fournier, Paola Antonelli, Kris Schenck, Hardy Kagimoto, Maggie Lewis, Jeff Le, Megan Palmer, Andrea Wong, and Matt Chessen who generously created opportunities for me to learn about the issues we present in this book: artificial intelligence, biotechnology, warfare, geopolitics, the global economy, global supply chains, and decision making at the highest levels of the US government, as well as synthetic biology's ethical challenges, geopolitical implications, and economic opportunities. Several of them patiently explained complicated facets of biology, read early portions of the manuscript, and introduced me to others working in the field.

I'm indebted to my partner and husband Dr. Brian Woolf, who listens to my hypotheses, reads first drafts, and challenges my ideas. I tested his patience regularly while writing this book as I asked him questions about academic research, talked him through the details of genetic editing, and debated endlessly about whether biological teleportation is possible (or whether I was just describing a high-tech fax machine). My dad, Don Webb, lived with us during the first year of the COVID-19 pandemic and read many early drafts, and my daughter, Petra, helped brainstorm the scenario chapters.

My incredible Spark Camp community listened to early ideas and concepts. In particular, Esther Dyson, was a source of continual inspiration. She reminds me to challenge my own cherished beliefs, encourages self-reflection, and energizes me to think bigger thoughts. James Geary and Ann Marie Lipinski at Harvard have been incredibly generous for many years, making it possible for me to host gatherings to talk about the future and to further develop my foresight methodology.

The US-Japan Leadership Program community, where I am a Fellow, is a supportive, brilliant group of people dedicated to building better futures. Kelly Nixon, James Ulak, George Packard, Tomoyuki Watanabe, and Aya Tsujita: I see the world differently because of your dedication and effort. Both *The Genesis Machine* and *The Big Nine* changed direction because of our Delegates weeks and Fellows weekends together, and because of ongoing conversations I've had with USJLPers.

The SynBioBeta community was incredibly welcoming, and I'm humbled by the innovators who gave me the opportunity to learn about their work. In the middle of the COVID-19 pandemic, SynBioBeta transitioned its annual conference to a virtual format in the fall of 2020, and it came just as we were starting to write. The online and offline conversations I had with speakers and participants proved invaluable. In addition, off-the-record conversations with my fellow members of the Council on Foreign Relations played an important role in my research.

I'm fortunate to do a lot of my strategic foresight thinking at New York University's Stern School of Business, where I research businesses and how they plan for the future. I'm grateful to Professor Sam Craig for bringing me into the MBA program and for advising me over the past few years. I cannot say enough about the incredibly bright, creative MBA students who have taken my classes. During the Fall 2020 and Spring 2021 semesters, I had the opportunity to test some synthetic biology scenarios during re-perception exercises with my students, who offered brilliant insights.

I am fortunate to know uber-connector Danny Stern, who insists that I think more exponentially. Mel Blake has mentored me, shaped my ideas, pushed me past my comfort zone, and called me to serve a greater purpose than the one I'd originally imagined for myself. He also introduced me to Andrew, for which I'll be forever grateful. Thanks to Mark Fortier and Lisa Barnes at Fortier Public Relations, and to Jamie Leifer and Miguel Cervantes at PublicAffairs whose patience seems to know no bounds, works to make sure my books are read by news media representatives and newsmakers alike and that I'm always prepared to have a meaningful conversation. My deepest appreciation to Clive Priddle for publishing my previous two books and for allowing me to explore challenging subjects. As ever, I'm grateful to Professor Sam Freedman, who started me on my book-writing journey while I was a graduate student at Columbia.

Without my Future Today Institute team, I would never have finished this book. Cheryl Cooney organized our client projects and workstreams to allow me to have the time I needed to research and write. My incredible colleague Emily Caufield managed the design and production of our annual trend report as I wrapped up the final chapters. Maureen Adams kept all of us on track, and me accountable during a very hectic year of virtual work.

Finally, I am indebted to Jon Fine, Carol Franco, Kent Lineback, and John Mahaney. Jon has edited me for many years—he knows my voice better than anyone (and he disapproves of that em dash I just used). Jon helped animate the stories in this book, made sure we offered color and detail, and reminded us to explain the science. Jon is a punk rock star (literally), a gifted editor, but importantly, a dear friend. Each time I begin a new project, my literary agent Carol and her husband Kent host me at their lovely home in Santa Fe to woodshed ideas, refine themes, and articulate the "promise" of the book. We spent days and nights distilling the research, concepts, characters, and ideas into core arguments, and in between work sessions we strolled around town and had lively discussions at terrific restaurants. Thanks to Carol, I met my editor John Mahaney, who has now shepherded three of my books. This one required a leap of faith and a considerable amount of trust, especially as initial deadlines passed. John, I've so enjoyed getting to know you these many years, and I still can't believe how fortunate I am that I get to work with you.

**Andrew**—So many people have been a part of the experiences distilled into this book that it is impossible to list all of them. The following have played key roles in my development, and I would like to extend my thanks to them: Betty McCaffrey, who made me, mostly; Frank Herbert, Arthur C. Clarke, James Cameron, Ridley Scott, Michael Crichton, and many others for their incredible stories; Dr. Ken Sanderson, for introducing me to the wonders of bacteria and genome mapping; Dr. Tak Mak and Amgen, for bringing me into the world of big science and big pharma; Drs. Craig Venter and Ham Smith, the dynamic duo of reading and writing genomes, for being so far ahead of the curve; Stephanie Selig, for showing a geeky scientist a completely different side to life, love, and spirituality; Dr. Tom Ray, for teaching me that the mind is as programmable as the cell; Drew Endy, Rob Carlson, Tom Knight, Randy Rettberg, Meagan Lizarazo, and others for creating the brilliant iGEM program and community.

My thanks also to Aubrey de Grey and Kevin Perrott for looping me into longevity and aging; Marc Hodosh for TEDMED; Drs. Chris Dambrowitz, Hans-Joachim Wieden, Christian Jacob, Michael Ellison, and many other folks in Alberta for leading the synbio community in Canada; John Carlson and Jason Tymko for exploring cooperative biotech with me; Peter Diamandis for Singularity University and X-Prize; everyone at Autodesk, especially Jonathan Knowles, Carl Bass, Jeff Kowalski, Carlos Olguin, and Larry Peck, for embracing biological CAD; Alicia Jackson for her wide-ranging mind; George Church, Jef Boeke, and Nancy Kelley for cofounding the Genome Project–Write, and Amy Schwartz for spearheading its growth; Jane Metcalfe for *Wired* and NEO.LIFE and long conversations.

Thanks as well to Rajeev Ronanki, Chad Moles, Peter Weijmarshausen, and 2048 Ventures, who brought Humane Genomics to life; Michael Hopmeier for his friendship, perspectives, and missile silo; NASA for daring mighty things; Elon Musk for doing the near-impossible again and again; Mickey McManus for being such an amazing all-around human being; and, of course, Hani, Ro, Dax, and the rest of the extended Hong family for unconditional love and the reason to build a better future. I'd clone each and every one of you, with permission. But *The Genesis Machine* would not exist without Amy's extensive research, writing, and mastery of the publication machine. Countless thanks to her and to Mel Blake and Danny Stern for connecting us. All first-time authors should be so lucky.

# *NOTES*

## Introduction: Should Life Be a Game of Chance?

1. Amy Webb, "All the Pregnancies I Couldn't Talk About," as first published in *The Atlantic*, October 21, 2019.

2. Heidi Ledford, "Five Big Mysteries About CRISPR's Origins," *Nature News* 541, no. 7637 (January 19, 2017): 280, https://doi.org/10.1038/541280a.

3. "Daily Updates of Totals by Week and State," Centers for Disease Control and Prevention, www.cdc.gov/nchs/nvss/vsrr/covid19/index.htm.

4. Julius Fredens, Kaihang Wang, Daniel de la Torre, Louise F. H. Funke, Wesley E. Robertson, Yonka Christova, Tiongsun Chia, et al., "Total Synthesis of *Escherichia coli* with a Recoded Genome," *Nature* 569, no. 7757 (May 1, 2019): 514–18, https://doi.org/10.1038/s41586-019-1192-5.

5. Embriette Hyde, "Why China Is Primed to Be the Ultimate SynBio Market," SynBioBeta, February 12, 2019, https://synbiobeta.com/why-china-is-primed-to-be-the-ultimate-synbio-market.

6. Thomas Hout and Pankaj Ghemawat, "China vs the World: Whose Technology Is It?," *Harvard Business Review*, December 1, 2010, https://hbr.org/2010/12/china-vs-the-world-whose-technology-is-it.

## 1 Saying No to Bad Genes: The Birth of the Genesis Machine

1. Video interview conducted by Amy Webb with Bill McBain on October 9, 2020.

2. Awad M. Ahmed, "History of Diabetes Mellitus," *Saudi Medical Journal* 23, no. 4 (April 2002): 373–78.

3. Jacob Roberts, "Sickening Sweet," Science History Institute, December 8, 2015, www.sciencehistory.org/distillations/sickening-sweet.

4. L. J. Dominguez and G. Licata. "The discovery of insulin: what really happened 80 years ago," *Annali Italiani di Medicina Interna* 16, no. 3 (September 2001): 155–62.

5. Robert D. Simoni, Robert L. Hill, and Martha Vaughan, "The Discovery of Insulin: The Work of Frederick Banting and Charles Best," *Journal of Biological Chemistry* 277, no. 26 (June 28, 2002): e1–2, https://doi.org/10.1016/S0021-9258(19)66673-1.

6. Simoni et al., "Discovery of Insulin."

7. "The Nobel Prize in Physiology or Medicine 1923," Nobel Prize, www.nobelprize.org/prizes/medicine/1923/summary.

8. "100 Years of Insulin," Eli Lilly and Company, www.lilly.com/discovery/100-years-of-insulin.

9. "Two Tons of Pig Parts: Making Insulin in the 1920s," National Museum of American History, November 1, 2013, https://americanhistory.si.edu/blog/2013/11/two-tons-of-pig-parts-making-insulin-in-the-1920s.html.

10. "Statistics About Diabetes," American Diabetes Association, www.diabetes.org/resources/statistics/statistics-about-diabetes.

11. "Eli Lilly Dies at 91," *New York Times*, January 25, 1977, www.nytimes.com/1977/01/25/archives/eli-lilly-dies-at-91-philanthropist-and-exhead-of-drug-company.html.

12. "Cloning Insulin," Genentech, April 7, 2016, www.gene.com/stories/cloning-insulin.

13. "Our Founders," Genentech, www.gene.com/about-us/leadership/our-founders.

14. Victor K. McElheny, "Technology: Making Human Hormones with Bacteria," *New York Times*, December 7, 1977, http://timesmachine.nytimes.com/timesmachine/1977/12/07/96407192.html.

15. Victor K. McElheny, "Coast Concern Plans Bacteria Use for Brain Hormone and Insulin," *New York Times*, December 2, 1977, www.nytimes.com/1977/12/02/archives/coast-concern-plans-bacteria-use-for-brain-hormone-and-insulin.html.

16. "Kleiner-Perkins and Genentech: When Venture Capital Met Science," https://store.hbr.org/product/kleiner-perkins-and-genentech-when-venture-capital-met-science/813102.

17. "Value of 1976 US Dollars Today—Inflation Calculator," https://www.inflationtool.com/us-dollar/1976-to-present-value?amount=1000000.

18. K. Itakura, T. Hirose, R. Crea, A. D. Riggs, H. L. Heyneker, F. Bolivar, and H. W. Boyer, "Expression in *Escherichia coli* of a Chemically Synthesized Gene for the Hormone Somatostatin," *Science* 198, no. 4321 (December 9, 1977): 1056–63, https://doi.org/10.1126/science.412251.

19. "Genentech," Kleiner Perkins, www.kleinerperkins.com/case-study/genentech.

20. "Cloning Insulin."

21. "Cloning Insulin."

22. Suzanne White Junod, "Celebrating a Milestone: FDA's Approval of First Genetically-Engineered Product," https://www.fda.gov/media/110447/download.

23. "An Estimation of the Number of Cells in the Human Body," *Annals of Human Biology*, https://informahealthcare.com/doi/abs/10.3109/03014460.2013.807878.

24. Christopher T. Walsh, Robert V. O'Brien, and Chaitan Khosla, "Nonproteinogenic Amino Acid Building Blocks for Nonribosomal Peptide and Hybrid Polyketide Scaffolds," *Angewandte Chemie* 52, no. 28 (July 8, 2013): 7098–124, https://doi.org/10.1002/anie.201208344.

25. Kavya Balaraman, "Fish Turn on Genes to Adapt to Climate Change," *Scientific American*, October 27, 2016, www.scientificamerican.com/article/fish-turn-on-genes-to-adapt-to-climate-change.

26. Ewen Callaway, "DeepMind's AI Predicts Structures for a Vast Trove of Proteins," *Nature News*, July 22, 2021, www.nature.com/articles/d41586-021-02025-4.

27. AlphaFold team, "A Solution to a 50-Year-Old Grand Challenge in Biology," DeepMind, November 30, 2020, https://deepmind.com/blog/article/alphafold-a-solution-to-a-50-year-old-grand-challenge-in-biology.

28. "Why Diabetes Patients Are Getting Insulin from Facebook," Science Friday, December 13, 2019, www.sciencefriday.com/segments/diabetes-insulin-facebook.

29. "Diabetic Buy Sell Trade Community," Facebook, www.facebook.com/groups/483202212435921.

30. Michael Fralick and Aaron S. Kesselheim, "The U.S. Insulin Crisis—Rationing a Lifesaving Medication Discovered in the 1920s," *New England Journal of Medicine* 381, no. 19 (November 7, 2019): 1793–95, https://doi.org/10.1056/NEJMp1909402.

31. "'The Absurdly High Cost of Insulin'—as High as $350 a Bottle, Often 2 Bottles per Month Needed by Diabetics," National AIDS Treatment Advocacy Project, www.natap.org/2019/HIV/052819_02.htm.

32. "Insulin Access and Affordability Working Group: Conclusions and Recommendations | Diabetes Care," accessed May 31, 2021, https://care.diabetesjournals.org/content/41/6/1299.

33. William T. Cefalu, Daniel E. Dawes, Gina Gavlak, Dana Goldman, William H. Herman, Karen Van Nuys, Alvin C. Powers, Simeon I. Taylor, and Alan L. Yatvin, on behalf of the Insulin Access and Affordability Working Group, "Insulin Access and Affordability Working Group: Conclusions and Recommendations," *Diabetes Care* 41, no. 6 (2018): 1299–1311, https://care.diabetesjournals.org/content/41/6/1299.

34. Briana Bierschbach, "What You Need to Know About the Insulin Debate at the Capitol," MPR News, August 16, 2019, www.mprnews.org/story/2019/08/16/what-you-need-to-know-about-the-insulin-debate-at-the-capitol.

35. Fralick and Kesselheim, "The U.S. Insulin Crisis."

36. Daniel G. Gibson, John I. Glass, Carole Lartigue, Vladimir N. Noskov, Ray-Yuan Chuang, Mikkel A. Algire, Gwynedd A. Benders, et al., "Creation of a Bacterial Cell Controlled by a Chemically Synthesized Genome," *Science* 329, no. 5987 (July 2, 2010): 52–56, https://doi.org/10.1126/science.1190719.

37. "No More Needles! Using Microbiome and Synthetic Biology Advances to Better Treat Type 1 Diabetes," J. Craig Venter Institute, March 25, 2019, www.jcvi.org/blog/no-more-needles-using-microbiome-and-synthetic-biology-advances-better-treat-type-1-diabetes.

38. Carl Zimmer, "Copyright Law Meets Synthetic Life Meets James Joyce," *National Geographic*, March 15, 2011, www.nationalgeographic.com/science/article/copyright-law-meets-synthetic-life-meets-james-joyce.

## 2 A Race to the Starting Line

1. "A Brief History of the Department of Energy," US Department of Energy, www.energy.gov/lm/doe-history/brief-history-department-energy.

2. Robert Cook-Deegan, "The Alta Summit, December 1984," *Genomics* 5 (October 1989): 661–63, archived at Human Genome Project Information Archive, 1990–2003, https://web.ornl.gov/sci/techresources/Human_Genome/project/alta.shtml.

3. Deegan, "The Alta Summit."

4. "Oral History Collection," National Human Genome Research Institute, www.genome.gov/leadership-initiatives/History-of-Genomics-Program/oral-history-collection.

5. "About the Human Genome Project," Human Genome Project Information Archive, 1990–2003, https://web.ornl.gov/sci/techresources/Human_Genome/project/index.shtml.

6. Institute of Medicine, Committee to Study Decision, Division of Health and Sciences Policy, *Biomedical Politics*, ed. Kathi Hanna (Washington, DC: National Academies Press, 1991).

7. "Human Genome Project Timeline of Events," National Human Genome Research Institute, www.genome.gov/human-genome-project/Timeline-of-Events.

8. "Human Genome Project Timeline of Events."

9. "Mills HS Presents Craig Venter, Ph.D.," Millbrae Community Television, 2017, https://mctv.tv/events/mills-hs-presents-craig-venter-ph-d.

10. Stephen Armstrong, "How Superstar Geneticist Craig Venter Stays Ahead in Science," *Wired UK*, June 9, 2017, www.wired.co.uk/article/craig-venter-synthetic-biology-success-tips.

11. Jason Schmidt, "The Genome Warrior," *New Yorker*, June 4, 2000, www.newyorker.com/magazine/2000/06/12/the-genome-warrior-2.

12. "Genetics and Genomics Timeline: 1991," Genome News Network, www.genomenewsnetwork.org/resources/timeline/1991_Venter.php.

13. Schmidt, "Genome Warrior."

14. At the time, there was no consensus on how many genes were in the human genome. Even as late as 2000, scientists were betting on the number, with the average estimate being around 62,500.

15. Douglas Birch, "Race for the Genome," *Baltimore Sun*, May 18, 1999.

16. John Crace, "Double Helix Trouble," *The Guardian*, October 16, 2007, www.theguardian.com/education/2007/oct/16/highereducation.research.

17. "Human Genome Project Budget," Human Genome Project Information Archive, 1990–2003, https://web.ornl.gov/sci/techresources/Human_Genome/project/budget.shtml.

18. "CPI Calculator by Country," Inflation Tool, www.inflationtool.com.

19. "Rosalind Franklin: A Crucial Contribution," reprinted from Ilona Miko and Lorrie LeJeune, eds., *Essentials of Genetics* (Cambridge, MA: NPG Education, 2009), Unit 1.3, *Nature Education*, www.nature.com/scitable/topicpage/rosalind-franklin-a-crucial-contribution-6538012.

20. James D. Watson, *The Double Helix: A Personal Account of the Discovery of the Structure of DNA* (London: Weidenfeld and Nicolson, 1981).

21. Julia Belluz, "DNA Scientist James Watson Has a Remarkably Long History of Sexist, Racist Public Comments," Vox, January 15, 2019, www.vox.com/2019/1/15/18182530/james-watson-racist.

22. Tom Abate, "Nobel Winner's Theories Raise Uproar in Berkeley: Geneticist's Views Strike Many as Racist, Sexist," SF Gate, November 13, 2000, www.sfgate.com/science/article/Nobel-Winner-s-Theories-Raise-Uproar-in-Berkeley-3236584.php.

23. Brandon Keim, "James Watson Suspended from Lab, but Not for Being a Sexist Hater of Fat People," *Wired*, October 2007, www.wired.com/2007/10/james-watson-su.

24. "James Watson: Scientist Loses Titles After Claims over Race," BBC News, January 13, 2019, www.bbc.com/news/world-us-canada-46856779.

25. John H. Richardson, "James Watson: What I've Learned," *Esquire*, October 19, 2007, www.esquire.com/features/what-ive-learned/ESQ0107jameswatson.

26. Belluz, "James Watson Has a Remarkably Long History."

27. Clive Cookson, "Gene Genies," *Financial Times*, October 19, 2007, www.ft.com/content/3cd61dbc-7b7d-11dc-8c53-0000779fd2ac.

28. J. Craig Venter, *A Life Decoded: My Genome, My Life* (New York: Viking, 2007).

29. L. Roberts, "Why Watson Quit as Project Head," *Science* 256, no. 5055 (April 17, 1992): 301–2, https://doi.org/10.1126/science.256.5055.301.

30. "Norman Schwarzkopf, U.S. Commander in Gulf War, Dies at 78," Reuters, December 28, 2012, www.reuters.com/news/picture/norman-schwarzkopf-us-commander-in-gulf-idUSBRE8BR01920121228.

31. Anjuli Sastry and Karen Grigsby Bates, "When LA Erupted in Anger: A Look Back at the Rodney King Riots," National Public Radio, April 26, 2017, www.npr.org/2017/04/26/524744989/when-la-erupted-in-anger-a-look-back-at-the-rodney-king-riots.

32. Schmidt, "Genome Warrior."

33. Leslie Roberts, "Scientists Voice Their Opposition," *Science* 256, no. 5061 (May 29, 1992): 1273ff, https://link.gale.com/apps/doc/A12358701/HRCA?sid=googleScholar&xid=72ac1090.

34. Schmidt, "The Genome Warrior."

35. Robert Sanders, "Decoding the Lowly Fruit Fly," *Berkeleyan*, February 3, 1999, www.berkeley.edu/news/berkeleyan/1999/0203/fly.html.

36. Nicholas J. Loman and Mark J. Pallen, "Twenty Years of Bacterial Genome Sequencing," *Nature Reviews Microbiology* 13, no. 12 (December 2015): 787–94, https://doi.org/10.1038/nrmicro3565.

37. "Genetics and Genomics Timeline: 1995," Genome News Network, www.genomenewsnetwork.org/resources/timeline/1995_Haemophilus.php.

38. Kate Reddington, Stefan Schwenk, Nina Tuite, Gareth Platt, Danesh Davar, Helena Coughlan, Yoann Personne, et al., "Comparison of Established Diagnostic Methodologies and a Novel Bacterial SmpB Real-Time PCR Assay for Specific Detection of *Haemophilus influenzae* Isolates Associated with Respiratory Tract Infections," *Journal of Clinical Microbiology* 53, no. 9 (September 2015): 2854–60, https://doi.org/10.1128/JCM.00777-15.

39. "Two Bacterial Genomes Sequenced," *Human Genome News* 7, no. 1 (May-June 1995), Human Genome Project Information Archive, 1990–2003, https://web.ornl.gov/sci/techresources/Human_Genome/publicat/hgn/v7n1/05microb.shtml.

40. H. O. Smith, J. F. Tomb, B. A. Dougherty, R. D. Fleischmann, and J. C. Venter, "Frequency and Distribution of DNA Uptake Signal Sequences in the Haemophilus Influenzae Rd Genome," *Science* 269, no. 5223 (July 28, 1995): 538–40, https://doi.org/10.1126/science.7542802.

41. Claire M. Fraser, Jeannine D. Gocayne, Owen White, Mark D. Adams, Rebecca A. Clayton, Robert D. Fleischmann, Carol J. Bult, et al., "The Minimal Gene Complement of Mycoplasma Genitalium," *Science* 270, no. 5235 (October 20, 1995): 397–404, https://doi.org/10.1126/science.270.5235.397.

42. "3700 DNA Analyzer," National Museum of American History, https://americanhistory.si.edu/collections/search/object/nmah_1297334.

43. Unknown to Dovichi, Hideki Kambara at Hitachi Corporation had developed similar technology at the same time. Applied Biosystems eventually licensed both technologies and worked with Hitachi to develop the device. In 2001, *Science* would call both researchers "unsung heroes" of the genome project.

44. Jim Kling, "Where the Future Went," *EMBO Reports* 6, no. 11 (November 2005): 1012–14, https://doi.org/10.1038/sj.embor.7400553.

45. Douglas Birch, "Race for the Genome," *Baltimore Sun*, May 18, 1999.

46. Nicholas Wade, "In Genome Race, Government Vows to Move Up Finish," *New York Times*, September 15, 1998, www.nytimes.com/1998/09/15/science/in-genome-race-government-vows-to-move-up-finish.html.

47. Lisa Belkin, "Splice Einstein and Sammy Glick. Add a Little Magellan," *New York Times*, August 23, 1998, www.nytimes.com/1998/08/23/magazine/splice-einstein-and-sammy-glick-add-a-little-magellan.html.

48. Schmidt, "Genome Warrior."

49. Douglas Birch, "Daring Sprint to the Summit. The Quest: A Determined Hamilton Smith Attempts to Scale a Scientific Pinnacle—and Reconcile with Fam-

ily," *Baltimore Sun*, April 13, 1999, www.baltimoresun.com/news/bs-xpm-1999-04-13-9904130335-story.html.

50. "Gene Firm Labelled a 'Con Job,'" BBC News, March 6, 2000, http://news.bbc.co.uk/2/hi/science/nature/667606.stm.

51. Mark D. Adams, Susan E. Celniker, Robert A. Holt, Cheryl A. Evans, Jeannine D. Gocayne, Peter G. Amanatides, Steven E. Scherer, et al., "The Genome Sequence of *Drosophila melanogaster*," *Science* 287, no. 5461 (March 24, 2000): 2185–95, https://doi.org/10.1126/science.287.5461.2185.

52. Nicholas Wade, "Rivals on Offensive as They Near Wire in Genome Race," *New York Times*, May 7, 2000, www.nytimes.com/2000/05/07/us/rivals-on-offensive-as-they-near-wire-in-genome-race.html.

53. Nicholas Wade, "Analysis of Human Genome Is Said to Be Completed," *New York Times*, April 7, 2000, https://archive.nytimes.com/www.nytimes.com/library/national/science/040700sci-human-genome.html.

54. Wade, "Analysis of Human Genome."

55. "Press Briefing by Dr. Neal Lane, Assistant to the President for Science and Technology; Dr. Frances Collins, Director of the National Human Genome Research Institute; Dr. Craig Venter, President and Chief Scientific Officer, Celera Genomics Corporation; and Dr. Ari Patrinos, Associate Director for Biological and Environmental Research, Department of Energy, on the Completion of the First Survey of the Entire Human Genome," White House Press Release, June 26, 2000, Human Genome Project Information Archive, 1990–2003, https://web.ornl.gov/sci/techresources/Human_Genome/project/clinton3.shtml.

56. "June 2000 White House Event," White House Press Release, June 26, 2000, National Human Genome Research Institute, www.genome.gov/10001356/june-2000-white-house-event.

57. "June 2000 White House Event."

58. "June 2000 White House Event."

59. Andrew Brown, "Has Venter Made Us Gods?," *The Guardian*, May 20, 2010, www.theguardian.com/commentisfree/andrewbrown/2010/may/20/craig-venter-life-god.

### 3 The Bricks of Life

1. "Marvin Minsky, Ph.D," Academy of Achievement, https://achievement.org/achiever/marvin-minsky-ph-d.

2. Martin Campbell-Kelly, "Marvin Minsky Obituary," *The Guardian*, February 3, 2016, www.theguardian.com/technology/2016/feb/03/marvin-minsky-obituary.

3. Jeremy Bernstein, "Marvin Minsky's Vision of the Future," *New Yorker*, December 6, 1981, www.newyorker.com/magazine/1981/12/14/a-i.

4. Amy Webb, *The Big Nine: How the Tech Titans and Their Thinking Machines Could Warp Humanity* (New York: PublicAffairs, 2019).

5. "HMS Beagle: Darwin's Trip Around the World," National Geographic Resource Library, n.d., www.nationalgeographic.org/maps/hms-beagle-darwins-trip-around-world.

6. Webb, *The Big Nine*.

7. "Tom Knight," Internet Archive Wayback Machine, http://web.archive.org/web/20040202103232/http://www.ai.mit.edu/people/tk/tk.html.

8. "Synthetic Biology, IGEM and Ginkgo Bioworks: Tom Knight's Journey," iGem Digest, 2018, https://blog.igem.org/blog/2018/12/4/tom-knight.

9. Sam Roberts, "Harold Morowitz, 88, Biophysicist, Dies; Tackled Enigmas Big and Small," *New York Times*, April 1, 2016, www.nytimes.com/2016/04/02/science/harold-morowitz-biophysicist-who-tackled-enigmas-big-and-small-dies-at-88.html.

10. Adam Bluestein, "Tom Knight, Godfather of Synthetic Biology, on How to Learn Something New," *Fast Company*, August 28, 2012, www.fastcompany.com/3000760/tom-knight-godfather-synthetic-biology-how-learn-something-new.

11. Bluestein, "Tom Knight, Godfather."

12. "Synthetic Biology, IGEM and Ginkgo Bioworks."

13. Roger Collis, "The Growing Threat of Malaria," *New York Times*, December 10, 1993, www.nytimes.com/1993/12/10/style/IHT-the-growing-threat-of-malaria.html.

14. Institute of Medicine, Committee on the Economics of Antimalarial Drugs, *Saving Lives, Buying Time: Economics of Malaria Drugs in an Age of Resistance*, eds. Kenneth J. Arrow, Claire Panosian, and Hellen Gelband (Washington, DC: National Academies Press, 2004).

15. Nicholas J. White, Tran T. Hien, and François H. Nosten, "A Brief History of Qinghaosu," *Trends in Parasitology* 31, no. 12 (December 2015): 607–10, https://doi.org/10.1016/j.pt.2015.10.010.

16. Eran Pichersky and Robert A. Raguso, "Why Do Plants Produce So Many Terpenoid Compounds?," *New Phytologist* 220, no. 3 (2018): 692–702, https://doi.org/10.1111/nph.14178.

17. Michael Specter, "A Life of Its Own," *New Yorker*, September 21, 2009, www.newyorker.com/magazine/2009/09/28/a-life-of-its-own.

18. Institute of Medicine, *Saving Lives, Buying Time*.

19. Ben Hammersley, "At Home with the DNA Hackers," *Wired UK*, October 8, 2009, www.wired.co.uk/article/at-home-with-the-dna-hackers.

20. Lynn Conway, "The M.I.T. 1978 MIT VLSI System Design Course," University of Michigan, accessed May 31, 2021, https://ai.eecs.umich.edu/people/conway/VLSI/MIT78/MIT78.html.

21. Oliver Morton, "Life, Reinvented," *Wired*, January 1, 2005, www.wired.com/2005/01/mit-3.

22. If you have kids who watch *Phineas & Ferb*, a "repressilator" is exactly the kind of fantastical machine Dr. Doofenshmirtz would have invented.

23. Drew Endy, Tom Knight, Gerald Sussman, and Randy Rettberg, "IAP 2003 Activity," IAP website hosted by MIT, last updated December 5, 2002, http://web.mit.edu/iap/www/iap03/searchiap/iap-4968.html.

24. "Synthetic Biology 1.0 SB 1.0," collaborative notes hosted at www.course hero.com/file/78510074/Sb10doc.

25. Vincent J J Martin, Douglas J. Pitera, Sydnor T. Withers, Jack D. Newman, and Jay D. Keasling, "Engineering a Mevalonate Pathway in *Escherichia coli* for Production of Terpenoids," *Nature Biotechnology* 21 (2003): 796–802, doi:10.1038/ nbt833.

26. Specter, "A Life of Its Own."

27. Ron Weiss, Joseph Jacobson, Paul Modrich, Jim Collins, George Church, Christina Smolke, Drew Endy, David Baker, and Jay Keasling, "Engineering Life: Building a FAB for Biology," *Scientific American*, June 2006, www.scientific american.com/article/engineering-life-building.

28. Richard Van Noorden, "Demand for Malaria Drug Soars," *Nature* 466, no. 7307 (August 2010): 672–73, https://doi.org/10.1038/466672a.

29. Daniel Grushkin, "The Rise and Fall of the Company That Was Going to Have Us All Using Biofuels," *Fast Company*, August 8, 2012, www.fastcompany .com/3000040/rise-and-fall-company-was-going-have-us-all-using-biofuels.

30. Grushkin, "The Rise and Fall of the Company."

31. Kevin Bullis, "Amyris Gives Up Making Biofuels: Update," *MIT Technology Review*, February 10, 2012, www.technologyreview.com/2012/02/10/20483 /amyris-gives-up-making-biofuels-update.

32. "Not Quite the Next Big Thing," Prism, February 2018, www.asee-prism .org/not-quite-the-next-big-thing.

33. James Hendler, "Avoiding Another AI Winter," *IEEE Intelligent Systems* 23, no. 2 (March 1, 2008): 2–4, https://doi.org/10.1109/MIS.2008.20.

## 4 God, a Church, and a (Mostly) Woolly Mammoth

1. Jill Lepore, "The Strange and Twisted Life of 'Frankenstein,'" *New Yorker*, February 5, 2018, www.newyorker.com/magazine/2018/02/12/the-strange-and -twisted-life-of-frankenstein.

2. Paul Russell and Anders Kraal, "Hume on Religion," in *The Stanford Encyclopedia of Philosophy*, ed. Edward N. Zalta, Stanford University, Spring 2020, https://plato.stanford.edu/archives/spr2020/entries/hume-religion.

3. "George Church," *Colbert Report*, season 9, episode 4, October 4, 2012 (video clip), Comedy Central, www.cc.com/video-clips/fkt99i/the-colbert-report -george-church.

4. "George Church," Oral History Collection, National Human Genome Research Institute, www.genome.gov/player/h5f7sh3K7L0/PL1ay9ko4A8sk0o9O -YhseFHzbU2I2HQQp.

5. "George Church," Oral History Collection.

6. Sharon Begley, "A Feature, Not a Bug: George Church Ascribes His Visionary Ideas to Narcolepsy," Stat News, June 8, 2017, www.statnews.com/2017/06/08 /george-church-narcolepsy.

7. Begley, "A Feature, Not a Bug."

8. Patricia Thomas, "DNA as Data," *Harvard Magazine*, January 1, 2004, www.harvardmagazine.com/2004/01/dna-as-data.html.

9. J. Tian, H. Gong, N. Sheng, X. Zhou, E. Gulari, X. Gao, G. Church, "Accurate Multiplex Gene Synthesis from Programmable DNA Microchips," *Nature*, December 23, 2004, 432(7020): 1050–54, doi: 10.1038/nature03151, PMID: 15616567.

10. Jin Billy Li, Yuan Gao, John Aach, Kun Zhang, Gregory V. Kryukov, Bin Xie, Annika Ahlford, et al., "Multiplex Padlock Targeted Sequencing Reveals Human Hypermutable CpG Variations," *Genome Research* 19, no. 9 (September 1, 2009): 1606–15, doi.org/10.1101/gr.092213.109.

11. Jon Cohen, "How the Battle Lines over CRISPR Were Drawn," *Science*, February 15, 2017, www.sciencemag.org/news/2017/02/how-battle-lines-over-crispr-were-drawn.

12. "The Nobel Prize in Chemistry 2020," Nobel Prize, www.nobelprize.org/prizes/chemistry/2020/summary.

13. Elizabeth Cooney, "George Church Salutes Fellow CRISPR Pioneers' Historic Nobel Win," Stat News, October 7, 2020, www.statnews.com/2020/10/07/a-terrific-choice-george-church-salutes-fellow-crispr-pioneers-historic-nobel-win.

14. "George M. Church, Ph.D., Co-Founder and Advisor," eGenesis, www.egenesisbio.com/portfolio-item/george-m-church.

15. Peter Miller, "George Church: The Future Without Limit," *National Geographic*, June 1, 2014, www.nationalgeographic.com/science/article/140602-george-church-innovation-biology-science-genetics-de-extinction.

16. Personal Genome Project website: https://www.personalgenomes.org/.

17. Blaine Bettinger, "Esther Dyson and the 'First 10,'" The Genetic Genealogist, July 27, 2007, https://thegeneticgenealogist.com/2007/07/27/esther-dyson-and-the-first-10/.

18. Amy Harmon, "6 Billion Bits of Data About Me, Me, Me!" *New York Times*, June 3, 2007, sec. Week in Review. https://www.nytimes.com/2007/06/03/weekinreview/03harm.html.

19. Bettinger, "Esther Dyson."

20. Stephen Pinker, "My Genome, My Self," *New York Times*, January 7, 2009, www.nytimes.com/2009/01/11/magazine/11Genome-t.html.

21. "The Life of Dolly," University of Edinburgh, https://dolly.roslin.ed.ac.uk/facts/the-life-of-dolly/index.html.

22. Charles Q. Choi, "First Extinct-Animal Clone Created," *National Geographic*, February 10, 2009, www.nationalgeographic.com/science/article/news-bucardo-pyrenean-ibex-deextinction-cloning.

23. Nicholas Wade, "The Woolly Mammoth's Last Stand," *New York Times*, March 2, 2017, www.nytimes.com/2017/03/02/science/woolly-mammoth-extinct-genetics.html.

24. David Biello, "3 Billion to Zero: What Happened to the Passenger Pigeon?," *Scientific American*, June 27, 2014, www.scientificamerican.com/article/3-billion-to-zero-what-happened-to-the-passenger-pigeon.

25. TEDx DeExtinction, https://reviverestore.org/events/tedxdeextinction.

26. "Hybridizing with Extinct Species: George Church at TEDx DeExtinction," www.youtube.com/watch?v=oTH_fmQo3Ok.

27. Christina Agapakis, "Alpha Males and Adventurous Human Females: Gender and Synthetic Genomics," *Scientific American*, January 22, 2013, https://blogs.scientificamerican.com/oscillator/alpha-males-and-adventurous-human-females-gender-and-synthetic-genomics.

28.George Church and coauthor Ed Regis described this scenario in the introduction of *Regenesis: How Synthetic Biology Will Reinvent Nature and Ourselves* (New York: Basic Books, 2014).

29. Gina Kolata, "Scientist Reports First Cloning Ever of Adult Mammal," *New York Times*, February 23, 1997, https://archive.nytimes.com/www.nytimes.com/books/97/12/28/home/022397clone-sci.html.

30. "Experts Detail Obstacles to Human Cloning," *MIT News*, May 14, 1997, https://news.mit.edu/1997/cloning-0514.

31. "Human Cloning: Ethical Issues," Church of Scotland, Church and Society Council, pamphlet, n.d., www.churchofscotland.org.uk/__data/assets/pdf_file/0006/3795/Human_Cloning_Ethical_Issues_leaflet.pdf.

32. "President Bill Clinton, March 4, 1997," transcript at CNN, www.cnn.com/ALLPOLITICS/1997/03/04/clinton.money/transcript.html.

33. "Poll: Most Americans Say Cloning Is Wrong," CNN.com, March 1, 1997, www.cnn.com/TECH/9703/01/clone.poll.

34. Editors, "Why Efforts to Bring Extinct Species Back from the Dead Miss the Point," *Scientific American*, June 1, 2013, www.scientificamerican.com/article/why-efforts-bring-extinct-species-back-from-dead-miss-point, https://doi.org/10.1038/scientificamerican0613-12.

35. George Church, "George Church: De-Extinction Is a Good Idea," *Scientific American*, September 1, 2013, www.scientificamerican.com/article/george-church-de-extinction-is-a-good-idea, https://doi.org/10.1038/scientificamerican0913-12.

36.TEDx DeExtinction, https://reviverestore.org/projects/woolly-mammoth/.

37. Ross Andersen, "Welcome to Pleistocene Park," *The Atlantic*, April 2017, www.theatlantic.com/magazine/archive/2017/04/pleistocene-park/517779.

38. Nathan Nunn and Nancy Qian, "The Columbian Exchange: A History of Disease, Food, and Ideas," *Journal of Economic Perspectives* 24, no. 2 (May 1, 2010): 163–88, https://doi.org/10.1257/jep.24.2.163.

39. Nunn and Qian, "The Columbian Exchange."

40. "The Human Cost of Disasters," UNDRR, October 12, 2020, https://reliefweb.int/report/world/human-cost-disasters-overview-last-20-years-2000-2019.

41. "The Human Cost of Disasters—An Overview of the Last 20 Years, 2000–2019," Relief Web, October 12, 2020, https://reliefweb.int/report/world/human-cost-disasters-overview-last-20-years-2000-2019.

42. Camilo Mora, Chelsie W. W. Counsell, Coral R. Bielecki, and Leo V Louis, "Twenty-Seven Ways a Heat Wave Can Kill You in the Era of Climate Change," *Circulation: Cardiovascular Quality and Outcomes* 10, no. 11 (November 1, 2017): e004233, https://doi.org/10.1161/CIRCOUTCOMES.117.004233.

43. "UN Report: Nature's Dangerous Decline 'Unprecedented'; Species Extinction Rates 'Accelerating,'" United Nations, Sustainable Development Goals, May 6, 2019, www.un.org/sustainabledevelopment/blog/2019/05/nature-decline-unprecedented-report.

44. Sinéad M. Crotty, Collin Ortals, Thomas M. Pettengill, Luming Shi, Maitane Olabarrieta, Matthew A. Joyce, and Andrew H. Altieri, "Sea-Level Rise and the Emergence of a Keystone Grazer Alter the Geomorphic Evolution and Ecology of Southeast US Salt Marshes," *Proceedings of the National Academy of Sciences* 117, no. 30 (July 28, 2020): 17891–902, www.pnas.org/content/117/30/17891.

45. "The Almond and Peach Trees Genomes Shed Light on the Differences Between These Close Species: Transposons Could Lie at the Origin of the Differences Between the Fruit of Both Species or the Flavor of the Almond," Science Daily, September 25, 2019, www.sciencedaily.com/releases/2019/09/190925123420.htm.

46. "President Obama Announces Intent to Nominate Francis Collins as NIH Director," White House Press Release, July 8, 2009, https://obamawhitehouse.archives.gov/the-press-office/president-obama-announces-intent-nominate-francis-collins-nih-director.

## 5 The Bioeconomy

1. Zhuang Pinghui, "Chinese Laboratory That First Shared Coronavirus Genome with World Ordered to Close for 'Rectification,' Hindering Its Covid-19 Research," *South China Morning Post*, February 28, 2020, www.scmp.com/news/china/society/article/3052966/chinese-laboratory-first-shared-coronavirus-genome-world-ordered.

2. Grady McGregor, "How an Overlooked Scientific Feat Led to the Rapid Development of COVID-19 Vaccines," *Fortune*, December 23, 2020, https://fortune.com/2020/12/23/how-an-overlooked-scientific-feat-led-to-the-rapid-development-of-covid-19-vaccines.

3. Yong-Zhen Zhang and Edward C. Holmes, "A Genomic Perspective on the Origin and Emergence of SARS-CoV-2," *Cell* 181, no. 2 (April 16, 2020): 223–27, https://doi.org/10.1016/j.cell.2020.03.035.

4. "Novel 2019 Coronavirus Genome," Virological, January 11, 2020, https://virological.org/t/novel-2019-coronavirus-genome/319.

5. "GenBank Overview," National Center for Biotechnology Information, www.ncbi.nlm.nih.gov/genbank.

6. "Novel 2019 Coronavirus Genome."

7. Walter Isaacson, "How mRNA Technology Could Upend the Drug Industry," *Time*, January 11, 2021, https://time.com/5927342/mrna-covid-vaccine.

8. Susie Neilson, Andrew Dunn, and Aria Bendix, "Moderna Groundbreaking Coronavirus Vaccine Was Designed in Just 2 Days," *Business Insider*, December 19, 2020, www.businessinsider.com/moderna-designed-coronavirus-vaccine-in-2-days-2020-11.

9. "The Speaking Telephone: Prof. Bell's Second Lecture Sending Multiple Dispatches in Different Directions over the Same Instrument at the Same Time Doing Away with Transmitters and Batteries a Substitute for a Musical Ear Autographs and Pictures By Telegraph," *New York Times*, May 19, 1877, www.nytimes.com/1877/05/19/archives/the-speaking-telephone-prof-bells-second-lecture-sending-multiple.html.

10. "The Speaking Telephone."

11. "AT&T's History of Invention and Breakups," *New York Times*, February 13, 2016, www.nytimes.com/interactive/2016/02/12/technology/att-history.html.

12. Arthur C. Clarke, "Extra-Terrestrial Relays: Can Rocket Stations Give World-Wide Radio Coverage?," In *Progress in Astronautics and Rocketry*, ed. Richard B. Marsten, 19: 3–6, Communication Satellite Systems Technology (Amsterdam: Elsevier, 1966), https://doi.org/10.1016/B978-1-4832-2716-0.50006-2.

13. Donald Martin, Paul Anderson, and Lucy Bartamian, "The History of Satellites," *Sat Magazine*, reprinted from *Communication Satellites*, 5th ed. (Reston, VA: American Institute of Aeronautics and Astronautics, 2007), www.satmagazine.com/story.php?number=768488682.

14. Mark Erickson, *Into the Unknown Together: The DOD, NASA, and Early Spaceflight* (Maxwell Air Force Base, AL: Air University Press, 2005).

15. As of this book's writing in 2021.

16. J. C. R. Licklider, "Memorandum for Members and Affiliates of the Intergalactic Computer Network," April 23, 1963, Advanced Research Projects Agency, archived at Metro Olografix, www.olografix.org/gubi/ estate/libri/wizards/memo.html.

17. Leonard Kleinrock, "The First Message Transmission," Internet Corporation for Assigned Names and Numbers (ICANN), October 29, 2019, www.icann.org/en/blogs/details/the-first-message-transmission-29-10-2019-en.

18. Ryan Singel, "Vint Cerf: We Knew What We Were Unleashing on the World," *Wired*, April 23, 2012, www.wired.com/2012/04/epicenter-isoc-famers-qa-cerf.

19. "History of the Web," World Wide Web Foundation, https://webfoundation.org/about/vision/history-of-the-web.

20. Sharita Forrest, "NCSA Web Browser 'Mosaic' Was Catalyst for Internet Growth," Illinois News Bureau, April 17, 2003, https://news.illinois.edu/view/6367/212344.

21. "Net Benefits," *The Economist*, March 9, 2013, www.economist.com/finance-and-economics/2013/03/09/net-benefits.

22. "U.S. Bioeconomy Is Strong, But Faces Challenges—Expanded Efforts in Coordination, Talent, Security, and Fundamental Research Are Needed," National

Academies of Sciences, Engineering, and Medicine, press release, January 14, 2020, www.nationalacademies.org/news/2020/01/us-bioeconomy-is-strong-but-faces-challenges-expanded-efforts-in-coordination-talent-security-and-fundamental-research-are-needed.

23. Michael Chui, Matthias Evers, James Manyika, Alice Zheng, and Travers Nisbet, "The Bio Revolution: Innovations Transforming Economies, Societies, and Our Lives," McKinsey and Company, May 13, 2020, www.mckinsey.com/industries/pharmaceuticals-and-medical-products/our-insights/the-bio-revolution-innovations-transforming-economies-societies-and-our-lives.

24. Stephanie Wisner, "Synthetic Biology Investment Reached a New Record of Nearly $8 Billion in 2020—What Does This Mean for 2021?," SynBioBeta, January 28, 2021, https://synbiobeta.com/synthetic-biology-investment-set-a-nearly-8-billion-record-in-2020-what-does-this-mean-for-2021.

25. Zhou Xin and Coco Feng, "ByteDance Value Approaches US$400 Billion as It Explores Douyin IPO," *South China Morning Post*, April 1, 2021, www.scmp.com/tech/big-tech/article/3128002/value-tiktok-maker-bytedance-approaches-us400-billion-new-investors.

26. Wisner, "Synthetic Biology Investment Reached a New Record."

27. "DNA Sequencing in Microgravity on the International Space Station (ISS) Using the MinION," Nanopore, August 29, 2016, https://nanoporetech.com/resource-centre/dna-sequencing-microgravity-international-space-station-iss-using-minion.

28. "Polynucleotide Synthesizer Model 280, Solid Phase Microprocessor Controller Model 100B," National Museum of American History, https://americanhistory.si.edu/collections/search/object/nmah_1451158.

29. US Security and Exchange Commission Form S-1/A filing by Twist Bioscience on October 17, 2018, SEC Archives, www.sec.gov/Archives/edgar/data/1581280/000119312518300580/d460243ds1a.htm.

30. "Building a Platform for Programming Technology," Microsoft Station B, https://www.microsoft.com/en-us/research/project/stationb.

31. Microsoft DNA Storage, https://www.microsoft.com/en-us/research/project/dna-storage.

32. "With a 'Hello,' Microsoft and UW Demonstrate First Fully Automated DNA Data Storage," Microsoft Innovation Stories, March 21, 2019, https://news.microsoft.com/innovation-stories/hello-data-dna-storage.

33. Robert F. Service, "DNA Could Store All of the World's Data in One Room," *Science*, March 2, 2017, www.sciencemag.org/news/2017/03/dna-could-store-all-worlds-data-one-room.

34. Nathan Hillson, Mark Caddick, Yizhi Cai, Jose A. Carrasco, Matthew Wook Chang, Natalie C. Curach, David J. Bell, et al., "Building a Global Alliance of Biofoundries," *Nature Communications* 10, no. 1 (May 9, 2019): 2040, https://doi.org/10.1038/s41467-019-10079-2.

35. "Moderna's Work on Our COVID-19 Vaccine," Moderna, www.modernatx.com/modernas-work-potential-vaccine-against-covid-19.

36. "Moderna's Work on Our COVID-19 Vaccine."

37. "'The Never Again Plan': Moderna CEO Stéphane Bancel Wants to Stop the Next Covid-19—Before It Happens," Advisory Board Company, December 22, 2020, www.advisory.com/Blog/2020/12/moderna-ceo-covid-vaccine-bancel.

38. Jacob Knutson, "Baltimore Plant Ruins 15 Million Johnson & Johnson Coronavirus Vaccines," Axios, March 31, 2021, www.axios.com/emergent-biosolutions-johnson-and-johnson-vaccine-dfd781a8-d007-4354-910a-e30d5007839b.html.

39. Jinshan Hong, Chloe Lo, and Michelle Fay Cortez, "Hong Kong Suspends BioNTech Shot over Loose Vial Caps, Stains," Bloomberg, March 24, 2021, www.bloomberg.com/news/articles/2021-03-24/macau-halts-biontech-shots-on-vials-hong-kong-rollout-disrupted.

40. Beatriz Horta, "Yale Lab Develops Revolutionary RNA Vaccine for Malaria," *Yale Daily News*, March 12, 2021, https://yaledailynews.com/blog/2021/03/12/yale-lab-develops-revolutionary-rna-vaccine-for-malaria.

41. Gordon E. Moore, "Cramming More Components onto Integrated Circuits, Reprinted from Electronics," *IEEE Solid-State Circuits Society Newsletter* 11, no. 3 (September 2006): 33–35, https://doi.org/10.1109/N-SSC.2006.4785860.

42. "The Cost of Sequencing a Human Genome," National Human Genome Research Institute, www.genome.gov/about-genomics/fact-sheets/Sequencing-Human-Genome-cost.

43. Antonio Regalado, "China's BGI Says It Can Sequence a Genome for Just $100," *MIT Technology Review*, February 26, 2020, www.technologyreview.com/2020/02/26/905658/china-bgi-100-dollar-genome.

44. Brian Alexander, "Biological Teleporter Could Seed Life Through Galaxy," *MIT Technology Review*, August 2, 2017, www.technologyreview.com/2017/08/02/150190/biological-teleporter-could-seed-life-through-galaxy.

## 6 The Biological Age

1. As told to Amy Webb in a video interview on September 24, 2020.

2. Philippa Roxby, "Malaria Vaccine Hailed as Potential Breakthrough," BBC News, April 23, 2021, www.bbc.com/news/health-56858158.

3. Hayley Dunning, "Malaria Mosquitoes Eliminated in Lab by Creating All-Male Populations," Imperial College London, News, May 11, 2020, www.imperial.ac.uk/news/197394/malaria-mosquitoes-eliminated-creating-all-male-populations.

4. "Scientists Release Controversial Genetically Modified Mosquitoes in High-Security Lab," National Public Radio, www.npr.org/sections/goatsandsoda/2019/02/20/693735499/scientists-release-controversial-genetically-modified-mosquitoes-in-high-securit.

5. "Landmark Project to Control Disease Carrying Mosquitoes Kicks Off in the Florida Keys," Cision, April 29, 2021, www.prnewswire.com/news-releases/landmark-project-to-control-disease-carrying-mosquitoes-kicks-off-in-the-florida-keys-301280593.html.

6. Lindsay Brownell, "Human Organ Chips Enable Rapid Drug Repurposing for COVID-19," Wyss Institute, May 3, 2021, https://wyss.harvard.edu/news/human-organ-chips-enable-rapid-drug-repurposing-for-covid-19.

7. "Body on a Chip," Wake Forest School of Medicine, https://school.wakehealth.edu/Research/Institutes-and-Centers/Wake-Forest-Institute-for-Regenerative-Medicine/Research/Military-Applications/Body-on-A-Chip.

8. Cleber A. Trujillo and Alysson R. Muotri, "Brain Organoids and the Study of Neurodevelopment," *Trends in Molecular Medicine* 24, no. 12 (December 2018): 982–90, https://doi.org/10.1016/j.molmed.2018.09.005.

9. "Stanford Scientists Assemble Human Nerve Circuit Driving Voluntary Movement," Stanford Medicine News Center, December 16, 2020, http://med.stanford.edu/news/all-news/2020/12/scientists-assemble-human-nerve-circuit-driving-muscle-movement.html.

10. "DeCODE Launches DeCODEme™," DeCODE Genetics, www.decode.com/decode-launches-decodeme.

11. Thomas Goetz, "23AndMe Will Decode Your DNA for $1,000. Welcome to the Age of Genomics," *Wired*, November 17, 2007, www.wired.com/2007/11/ff-genomics.

12. "23andMe Genetic Service Now Fully Accessible to Customers in New York and Maryland," 23andMe, December 4, 2015, https://mediacenter.23andme.com/press-releases/23andme-genetic-service-now-fully-accessible-to-customers-in-new-york-and-maryland.

13. "'Smart Toilet' Monitors for Signs of Disease," Stanford Medicine News Center, April 6, 2020, http://med.stanford.edu/news/all-news/2020/04/smart-toilet-monitors-for-signs-of-disease.html.

14. Mark Mimee, Phillip Nadeau, Alison Hayward, Sean Carim, Sarah Flanagan, Logan Jerger, Joy Collins, et al., "An Ingestible Bacterial-Electronic System to Monitor Gastrointestinal Health," *Science* 360, no. 6391 (May 25, 2018): 915–18, https://doi.org/10.1126/science.aas9315.

15. Tori Marsh, "Live Updates: January 2021 Drug Price Hikes," GoodRx, January 19, 2021, www.goodrx.com/blog/january-drug-price-hikes-2021.

16. "2019 Employer Health Benefits Survey. Section 1: Cost of Health Insurance," Kaiser Family Foundation, September 25, 2019, www.kff.org/report-section/ehbs-2019-section-1-cost-of-health-insurance.

17. Bruce Budowle and Angela van Daal, "Forensically Relevant SNP Classes," *BioTechniques* 44, no. 5 (April 1, 2008): 603–10, https://doi.org/10.2144/000112806.

18. Leslie A. Pray, "Embryo Screening and the Ethics of Human Genetic Engineering," *Nature Education* 1, no. 1 (2008): 207, www.nature.com/scitable/topicpage/embryo-screening-and-the-ethics-of-human-60561.

19. Antonio Regalado, "Engineering the Perfect Baby," *MIT Technology Review*, March 5, 2015, www.technologyreview.com/2015/03/05/249167/engineering-the-perfect-baby.

20. Rachel Lehmann-Haupt, "Get Ready for Same-Sex Reproduction," NEO.LIFE, February 28, 2018, https://neo.life/2018/02/get-ready-for-same-sex-reproduction.

21. Daisy A. Robinton and George Q Daley, "The Promise of Induced Pluripotent Stem Cells in Research and Therapy," *Nature* 481, no. 7381 (January 18, 2012): 295-305, doi:10.1038/nature10761.

22. "'Artificial Womb' Invented at the Children's Hospital of Philadelphia," WHYY PBS, April 25, 2017, https://whyy.org/articles/artificial-womb-invented-at-the-childrens-hospital-of-philadelphia.

23. Antonio Regalado, "A Mouse Embryo Has Been Grown in an Artificial Womb—Humans Could Be Next," *MIT Technology Review*, March 17, 2021, www.technologyreview.com/2021/03/17/1020969/mouse-embryo-grown-in-a-jar-humans-next.

24. "Our Current Water Supply," Southern Nevada Water Authority, https://www.snwa.com/water-resources/current-water-supply/index.html.

25. "Food Loss and Waste Database," United Nations, Food and Agriculture Organization, www.fao.org/food-loss-and-food-waste/flw-data.

26. "Sustainable Management of Food Basics," US Environmental Protection Agency, August 11, 2015, www.epa.gov/sustainable-management-food/sustainable-management-food-basics.

27. "Worldwide Food Waste," Think Eat Save, United Nations Environment Programme, www.unep.org/thinkeatsave/get-informed/worldwide-food-waste.

28. Kenneth A. Barton, Andrew N. Binns, Antonius J.M. Matzke, and Mary-Dell Chilton, "Regeneration of Intact Tobacco Plants Containing Full Length Copies of Genetically Engineered T-DNA, and Transmission of T-DNA to R1 Progeny," *Cell* 32, no. 4 (April 1, 1983): 1033–43, https://doi.org/10.1016/0092-8674(83)90288-X.

29. "Tremors in the Hothouse," *New Yorker*, July 19, 1993, www.newyorker.com/magazine/1993/07/19/tremors-in-the-hothouse.

30. "ISAAA Brief 55-2019: Executive Summary: Biotech Crops Drive Socio-Economic Development and Sustainable Environment in the New Frontier," International Service for the Acquisition of Agri-biotech Applications, 2019, www.isaaa.org/resources/publications/briefs/55/executivesummary/default.asp.

31. "Recent Trends in GE Adoption," US Department of Agriculture Economic Research Service, www.ers.usda.gov/data-products/adoption-of-genetically-engineered-crops-in-the-us/recent-trends-in-ge-adoption.aspx.

32. Javier Garcia Martinez, "Artificial Leaf Turns Carbon Dioxide into Liquid Fuel," *Scientific American*, June 26, 2017, www.scientificamerican.com/article/liquid-fuels-from-sunshine.

33. Max Roser and Hannah Ritchie, "Hunger and Undernourishment," Our World in Data, October 8, 2019, https://ourworldindata.org/hunger-and-undernourishment.

34. "Growing at a Slower Pace, World Population Is Expected to Reach 9.7 Billion in 2050 and Could Peak at Nearly 11 Billion Around 2100," United Nations, Department of Economic and Social Affairs, June 17, 2019, www.un.org/development/desa/en/news/population/world-population-prospects-2019.html.

35. Julia Moskin, Brad Plumer, Rebecca Lieberman, Eden Weingart, and Nadja Popovich, "Your Questions About Food and Climate Change, Answered," *New York Times*, April 30, 2019, www.nytimes.com/interactive/2019/04/30/dining/climate-change-food-eating-habits.html.

36. "China's Breeding Giant Pigs That Are as Heavy as Polar Bears," Bloomberg, October 6, 2019, www.bloomberg.com/news/articles/2019-10-06/china-is-breeding-giant-pigs-the-size-of-polar-bears.

37. Kristine Servando, "China's Mutant Pigs Could Help Save Nation from Pork Apocalypse," Bloomberg, December 3, 2019, www.bloomberg.com/news/features/2019-12-03/china-and-the-u-s-are-racing-to-create-a-super-pig.

38. "Belgian Blue," The Cattle Site, www.thecattlesite.com/breeds/beef/8/belgian-blue.

39. Antonio Regalado, "First Gene-Edited Dogs Reported in China," *MIT Technology Review*, October 19, 2015, www.technologyreview.com/2015/10/19/165740/first-gene-edited-dogs-reported-in-china.

40. Robin Harding, "Vertical Farming Finally Grows Up in Japan," *Financial Times*, January 22, 2020, www.ft.com/content/f80ea9d0-21a8-11ea-b8a1-584213ee7b2b.

41. Winston Churchill, "Fifty Years Hence," *Maclean's*, November 15, 1931, https://archive.macleans.ca/article/1931/11/15/fifty-years-hence.

42. Alok Jha, "World's First Synthetic Hamburger Gets Full Marks for 'Mouth Feel,'" *The Guardian*, August 6, 2013, www.theguardian.com/science/2013/aug/05/world-first-synthetic-hamburger-mouth-feel.

43. Bec Crew, "Cost of Lab-Grown Burger Patty Drops from $325,000 to $11.36," Science Alert, April 2, 2015, www.sciencealert.com/lab-grown-burger-patty-cost-drops-from-325-000-to-12.

44. Karen Gilchrist, "This Multibillion-Dollar Company Is Selling Lab-Grown Chicken in a World-First," CNBC, March 1, 2021, www.cnbc.com/2021/03/01/eat-just-good-meat-sells-lab-grown-cultured-chicken-in-world-first.html.

45. Kai Kupferschmidt, "Here It Comes . . . The $375,000 Lab-Grown Beef Burger," *Science*, August 2, 2013, www.sciencemag.org/news/2013/08/here-it-comes-375000-lab-grown-beef-burger.

46. "WHO's First Ever Global Estimates of Foodborne Diseases Find Children Under 5 Account for Almost One Third of Deaths," World Health Organization, December 3, 2015, www.who.int/news/item/03-12-2015-who-s-first-ever-global-estimates-of-foodborne-diseases-find-children-under-5-account-for-almost-one-third-of-deaths.

47. "Outbreak of *E. coli* Infections Linked to Romaine Lettuce," Centers for Disease Control and Prevention, January 15, 2020, www.cdc.gov/ecoli/2019/o157h7-11-19/index.html.

48. Kevin Jiang, "Synthetic Microbial System Developed to Find Objects' Origin," *Harvard Gazette*, June 4, 2020, https://news.harvard.edu/gazette/story/2020/06/synthetic-microbial-system-developed-to-find-objects-origin.

49. Jen Alic, "Is the Future of Biofuels in Algae? Exxon Mobil Says It's Possible," *Christian Science Monitor*, March 13, 2013, www.csmonitor.com/Environment/Energy-Voices/2013/0313/Is-the-future-of-biofuels-in-algae-Exxon-Mobil-says-it-s-possible.

50. "J. Craig Venter Institute–Led Team Awarded 5-Year, $10.7 M Grant from US Department of Energy to Optimize Metabolic Networks in Diatoms, Enabling

Next-Generation Biofuels and Bioproducts," J. Craig Venter Institute, October 3, 2017, www.jcvi.org/media-center/j-craig-venter-institute-led-team-awarded-5-year-107-m-grant-us-department-energy.

51. "Advanced Algal Systems," US Department of Energy, www.energy.gov/eere/bioenergy/advanced-algal-systems.

52. Morgan McFall-Johnsen, "These Facts Show How Unsustainable the Fashion Industry Is," World Economic Forum, January 31, 2020, www.weforum.org/agenda/2020/01/fashion-industry-carbon-unsustainable-environment-pollution.

53. Rachel Cormack, "Why Hermès, Famed for Its Leather, Is Rolling Out a Travel Bag Made from Mushrooms," *Robb Report*, March 15, 2021, https://robbreport.com/style/accessories/hermes-vegan-mushroom-leather-1234601607.

54. "Genomatica to Scale Bio-Nylon 50-Fold with Aquafil," Genomatica, press release, November 19, 2020, www.genomatica.com/bio-nylon-scaling-50x-to-support-global-brands.

55. L. Lebreton, B. Slat, F. Ferrari, B. Sainte-Rose, J. Aitken, R. Marthouse, S. Hajbane, et al., "Evidence That the Great Pacific Garbage Patch Is Rapidly Accumulating Plastic," *Scientific Reports* 8, no. 1 (March 22, 2018): 4666, https://doi.org/10.1038/s41598-018-22939-w.

56. "Ocean Trash: 5.25 Trillion Pieces and Counting, but Big Questions Remain," National Geographic Resource Library, n.d., www.nationalgeographic.org/article/ocean-trash-525-trillion-pieces-and-counting-big-questions-remain/6th-grade.

## 7 Nine Risks

1. Emily Waltz, "Gene-Edited CRISPR Mushroom Escapes U.S. Regulation: Nature News and Comment," *Nature* 532, no. 293 (2016), www.nature.com/news/gene-edited-crispr-mushroom-escapes-us-regulation-1.19754.

2. Waltz, "Gene-Edited CRISPR Mushroom."

3. Antonio Regalado, "Here Come the Unregulated GMOs," *MIT Technology Review*, April 15, 2016, www.technologyreview.com/2016/04/15/8583/here-come-the-unregulated-gmos.

4. Waltz, "Gene-Edited CRISPR Mushroom."

5. Doug Bolton, "Mushrooms that don't turn brown could soon be on sale thanks to loophole in GM food regulations," *The Independent*, April 18, 2016, https://www.independent.co.uk/news/science/gene-editing-mushrooms-usda-regulations-approved-edited-brown-a6989531.html.

6. "如果你不能接受转基因，基因编辑食品你敢吃吗？|转基因|基因编辑|食物_新浪科技_新浪网," Sina Technology, June 30, 2016, http://tech.sina.com.cn/d/i/2016-06-30/doc-ifxtsatn7803705.shtml.

7. Andrew MacFarlane, "Genetically Modified Mushrooms May Lead the Charge to Ending World Hunger," Weather Channel, April 20, 2016, https://weather.com/science/news/genetically-modified-mushrooms-usda.

8. "Secretary Perdue Issues USDA Statement on Plant Breeding Innovation," US Department of Agriculture, Animal and Plant Health Inspection Service, March 28, 2018, https://content.govdelivery.com/accounts/USDAAPHIS/bulletins/1e599ff.

9. Pam Belluck, "Chinese Scientist Who Says He Edited Babies' Genes Defends His Work," *New York Times*, November 28, 2018, www.nytimes.com/2018/11/28/world/asia/gene-editing-babies-he-jiankui.html.

10. Belluck, "Chinese Scientist."

11. "He Jiankui's Gene Editing Experiment Ignored Other HIV Strains," Stat News, April 15, 2019, www.statnews.com/2019/04/15/jiankui-embryo-editing-ccr5.

12. Antonio Regalado, "China's CRISPR Twins Might Have Had Their Brains Inadvertently Enhanced," *MIT Technology Review*, February 21, 2019, www.technologyreview.com/2019/02/21/137309/the-crispr-twins-had-their-brains-altered.

13. For the original agenda, see "Second International Summit on Human Gene Editing," National Academies of Sciences, Engineering, and Medicine, November 27, 2018, www.nationalacademies.org/event/11-27-2018/second-international-summit-on-human-gene-editing.

14. David Cyranoski, "What CRISPR-Baby Prison Sentences Mean for Research," *Nature* 577, no. 7789 (January 3, 2020): 154–55, https://doi.org/10.1038/d41586-020-00001-y.

15. Anders Lundgren, "Carl Wilhelm Scheele: Swedish Chemist," Encyclopedia Britannica, www.britannica.com/biography/Carl-Wilhelm-Scheele.

16. Gilbert King, "Fritz Haber's Experiments in Life and Death," *Smithsonian Magazine*, June 6, 2012, www.smithsonianmag.com/history/fritz-habers-experiments-in-life-and-death-114161301.

17. Jennifer Couzin-Frankel, "Poliovirus Baked from Scratch," *Science*, July 11, 2002, www.sciencemag.org/news/2002/07/poliovirus-baked-scratch.

18. "Traces of Terror. The Science: Scientists Create a Live Polio Virus," *New York Times*, July 12, 2002, www.nytimes.com/2002/07/12/us/traces-of-terror-the-science-scientists-create-a-live-polio-virus.html.

19. Kai Kupferschmidt, "How Canadian Researchers Reconstituted an Extinct Poxvirus for $100,000 Using Mail-Order DNA," *Science*, July 6, 2017, www.sciencemag.org/news/2017/07/how-canadian-researchers-reconstituted-extinct-poxvirus-100000-using-mail-order-dna.

20. Denise Grady and Donald G. McNeil Jr., "Debate Persists on Deadly Flu Made Airborne," *New York Times*, December 27, 2011, www.nytimes.com/2011/12/27/science/debate-persists-on-deadly-flu-made-airborne.html.

21. Monica Rimmer, "How Smallpox Claimed Its Final Victim," BBC News, August 10, 2018, www.bbc.com/news/uk-england-birmingham-45101091.

22. J. Kenneth Wickiser, Kevin J. O'Donovan, Michael Washington, Stephen Hummel, and F. John Burpo, "Engineered Pathogens and Unnatural Biological Weapons: The Future Threat of Synthetic Biology," *CTC Sentinel* 13, no. 8 (August 31, 2020): 1–7, https://ctc.usma.edu/engineered-pathogens-and-unnatural-biological-weapons-the-future-threat-of-synthetic-biology.

23. Ian Sample, "Craig Venter Creates Synthetic Life Form," *The Guardian*, May 20, 2010, www.theguardian.com/science/2010/may/20/craig-venter-synthetic-life-form.

24. Margaret Munro, "Life, From Four Chemicals," *Ottawa Citizen*, May 21, 2010, www.pressreader.com/canada/ottawa-citizen/20100521/285121404908322.

25. Sample, "Craig Venter Creates Synthetic Life Form."

26. Ian Sample, "Synthetic Life Breakthrough Could Be Worth over a Trillion Dollars," *The Guardian*, May 20, 2010, www.theguardian.com/science/2010/may/20/craig-venter-synthetic-life-genome.

27. Clyde A. Hutchison, Ray-Yuan Chuang, Vladimir N. Noskov, Nacyra Assad-Garcia, Thomas J. Deerinck, Mark H. Ellisman, John Gill, et al., "Design and Synthesis of a Minimal Bacterial Genome," *Science* 351, no. 6280 (March 25, 2016), https://doi.org/10.1126/science.aad6253.

28. "Scientists Create Simple Synthetic Cell That Grows and Divides Normally," National Institute of Standards and Technology, March 29, 2021, www.nist.gov/news-events/news/2021/03/scientists-create-simple-synthetic-cell-grows-and-divides-normally.

29. Ken Kingery, "Engineered Swarmbots Rely on Peers for Survival," Duke Pratt School of Engineering, February 29, 2016, https://pratt.duke.edu/about/news/engineered-swarmbots-rely-peers-survival.

30. Rob Stein, "Blind Patients Hope Landmark Gene-Editing Experiment Will Restore Their Vision," National Public Radio, May 10, 2021, www.npr.org/sections/health-shots/2021/05/10/993656603/blind-patients-hope-landmark-gene-editing-experiment-will-restore-their-vision.

31. Sigal Samuel, "A Celebrity Biohacker Who Sells DIY Gene-Editing Kits Is Under Investigation," Vox, May 19, 2019, www.vox.com/future-perfect/2019/5/19/18629771/biohacking-josiah-zayner-genetic-engineering-crispr.

32. Arielle Duhaime-Ross, "In Search of a Healthy Gut, One Man Turned to an Extreme DIY Fecal Transplant," The Verge, May 4, 2016, www.theverge.com/2016/5/4/11581994/fmt-fecal-matter-transplant-josiah-zayner-microbiome-ibs-c-diff.

33. Stephanie M. Lee, "This Biohacker Is Trying to Edit His Own DNA and Wants You to Join Him," BuzzFeed, October 14, 2017, www.buzzfeednews.com/article/stephaniemlee/this-biohacker-wants-to-edit-his-own-dna.

34. Molly Olmstead, "The Fuzzy Regulations Surrounding DIY Synthetic Biology," Slate, May 4, 2017, https://slate.com/technology/2017/05/the-fuzzy-regulations-surrounding-diy-synthetic-biology.html.

35. Doudna and Zheng each founded four, including Scribe Therapeutics, Intellia Therapeutics, Mammoth Biosciences, and Caribou Biosciences (Doudna) and Sherlock Biosciences, Arbor Biotechnologies, Beam Therapeutics, and Editas Medicine (Zheng). Charpentier founded two: CRISPR Therapeutics and ERS Genomics. Doudna was an original cofounder of Editas, but broke ties with Zheng over the patent dispute.

36. "Statement from Ambassador Katherine Tai on the Covid-19 Trips Waiver," Office of the United States Trade Representative, May 5, 2021, https://ustr.gov/about-us/policy-offices/press-office/press-releases/2021/may/statement-ambassador-katherine-tai-covid-19-trips-waiver.

37. Kate Taylor, "More Parents Plead Guilty in College Admissions Scandal," *New York Times*, October 21, 2019, www.nytimes.com/2019/10/21/us/college-admissions-scandal.html.

38. Andrew Martinez, "Lawyer Who Paid $75G to Fix Daughter's Test Answers Gets One-Month Prison Term," *Boston Herald*, October 3, 2019, www.bostonherald.com/2019/10/03/lawyer-who-paid-75g-to-fix-daughters-test-answers-gets-one-month-prison-term.

39. Matthew Campbell and Doug Lyu, "China's Genetics Giant Wants to Tailor Medicine to Your DNA," Bloomberg, November 13, 2019, www.bloomberg.com/news/features/2019-11-13/chinese-genetics-giant-bgi-wants-to-tailor-medicine-to-your-dna.

40. "China: Minority Region Collects DNA from Millions," Human Rights Watch, December 13, 2017, www.hrw.org/news/2017/12/13/china-minority-region-collects-dna-millions.

41. Sui-Lee Wee, "China Uses DNA to Track Its People, with the Help of American Expertise," *New York Times*, February 21, 2019, www.nytimes.com/2019/02/21/business/china-xinjiang-uighur-dna-thermo-fisher.html.

42. "China's Ethnic Tinderbox," BBC, July 9, 2009, http://news.bbc.co.uk/2/hi/asia-pacific/8141867.stm.

43. Simon Denyer, "Researchers May Have 'Found' Many of China's 30 Million Missing Girls," *Washington Post*, November 30, 2016, www.washingtonpost.com/news/worldviews/wp/2016/11/30/researchers-may-have-found-many-of-chinas-30-million-missing-girls.

44. Kirsty Needham, "Special Report: COVID Opens New Doors for China's Gene Giant," Reuters, August 5, 2020, www.reuters.com/article/us-health-coronavirus-bgi-specialreport-idUSKCN2511CE.

45. The Seasteading Institute hopes to create ocean-based communities outside the governing frameworks of established countries. Nobel Prize–winning economist Milton Friedman's grandson, Patri Friedman, and PayPal cofounder and venture capitalist Peter Thiel are cofounders. https://www.seasteading.org/.

46. "Todai-Led Team Creates Mouse Pancreas in Rat in Treatment Breakthrough," *Japan Times*, January 26, 2017, www.japantimes.co.jp/news/2017/01/26/national/science-health/treatment-breakthrough-todai-led-team-creates-mouse-pancreas-rat-transplants-diabetic-mouse.

47. Nidhi Subbaraman, "First Monkey–Human Embryos Reignite Debate over Hybrid Animals," *Nature* 592, no. 7855 (April 15, 2021): 497, https://doi.org/10.1038/d41586-021-01001-2.

48. Julian Savulescu and César Palacios-González, "First Human–Monkey Embryos Created—A Small Step Towards a Huge Ethical Problem," The Conversation, April 22, 2021, https://theconversation.com/first-human-monkey-embryos-created-a-small-step-towards-a-huge-ethical-problem-159355.

49. Alex Fox, "Compared with Hummingbirds, People Are Rather Colorblind," *Smithsonian Magazine*, June 18, 2020, www.smithsonianmag.com/smart-news/compared-hummingbirds-were-all-colorblind-180975111.

50. Guy Rosen, "How We're Tackling Misinformation Across Our Apps," Facebook, March 22, 2021, https://about.fb.com/news/2021/03/how-were-tackling-misinformation-across-our-apps.

51. Rosen, "How We're Tackling Misinformation."

52. Fortune 500, https://fortune.com/fortune500.

53. Healthy and Natural World Facebook Page, "Scientists Warn People to Stop Eating Instant Noodles Due to Cancer and Stroke Risks," Facebook.com, March 20, 2019, www.facebook.com/HealthyAndNaturalWorld/posts/scientists-warn-people-to-stop-eating-instant-noodles-due-to-cancer-and-stroke-r/2262994090426410.

54. Michelle R. Smith and Johnathan Reiss, "Inside One Network Cashing In on Vaccine Disinformation," Associated Press, May 13, 2021, https://apnews.com/article/anti-vaccine-bollinger-coronavirus-disinformation-a7b8e1f33990670563b4c469b462c9bf.

55. Smith and Reiss, "Inside One Network."

56. Ben Guarino, Ariana Eunjung Cha, Josh Wood, and Griff Witte, "'The Weapon That Will End the War': First Coronavirus Vaccine Shots Given Outside Trials in U.S.," December 14, 2020, www.washingtonpost.com/nation/2020/12/14/first-covid-vaccines-new-york.

57. "Coronavirus (COVID-19) Vaccinations," Our World In Data, https://ourworldindata.org/covid-vaccinations?country=USA.

58. "Provisional COVID-19 Death Counts by Week Ending Date and State," Centers for Disease Control and Prevention, https://data.cdc.gov/NCHS/Provisional-COVID-19-Death-Counts-by-Week-Ending-D/r8kw-7aab.

59. Jack Healy, "These Are the 5 People Who Died in the Capitol Riot," *New York Times*, January 11, 2021, https://www.nytimes.com/2021/01/11/us/who-died-in-capitol-building-attack.html.

60. "Public Trust in Government: 1958–2021," Pew Research Center, https://www.pewresearch.org/politics/2021/05/17/public-trust-in-government-1958-2021.

## 8 The Story of Golden Rice

1. Ian McNulty, "Next Generation to Reopen Li'l Dizzy's, Reviving New Orleans Restaurant Legacy," January 2, 2021, NOLA.com, www.nola.com/entertainment_life/eat-drink/article_a346001a-4d49-11eb-b927-a73cacd63596.html.

2. Confucius, *The Analects of Confucius*, trans. Arthur Waley (New York: Random House, 1989), Bk. 10.

3. Sarah Zhang, "Archaeologists Find Evidence of the First Rice Ever Grown," *The Atlantic*, May 29, 2017, www.theatlantic.com/science/archive/2017/05/rice-domestication/528288.

4. John Christensen, "Scientist at Work. Ingo Potrykus: Golden Rice in a Grenade-Proof Greenhouse," *New York Times*, November 21, 2000, www.nytimes.com/2000/11/21/science/scientist-at-work-ingo-potrykus-golden-rice-in-a-grenade-proof-greenhouse.html.

5. Interview with Dr. Brian Woolf by Amy Webb, August 15, 2020.

6. J. Madeleine Nash, "This Rice Could Save a Million Kids a Year," *Time*, July 31, 2000, http://content.time.com/time/magazine/article/0,9171,997586,00.html.

7. "The Rockefeller Foundation: A Long-Term Bet on Scientific Breakthrough," Rockefeller Foundation, https://engage.rockefellerfoundation.org/story-sketch/rice-biotechnology-research-network.

8. Christensen, "Scientist at Work."

9. Mary Lou Guerinot, "The Green Revolution Strikes Gold," *Science* 287, no. 5451 (January 14, 2000): 241–43, https://doi.org/10.1126/science.287.5451.241.

10. Nash, "This Rice Could Save a Million Kids."

11. David Barboza, "AstraZeneca to Sell a Genetically Engineered Strain of Rice," *New York Times*, May 16, 2000, www.nytimes.com/2000/05/16/business/astrazeneca-to-sell-a-genetically-engineered-strain-of-rice.html.

12. "GM Rice Patents Given Away," BBC News, August 4, 2000, http://news.bbc.co.uk/2/hi/science/nature/865946.stm.

13. Margaret Wertheim, "Frankenfoods," *LA Weekly*, July 5, 2000, www.laweekly.com/frankenfoods.

14. "Monsanto Pushes 'Golden Rice,'" CBS News, August 4, 2000, www.cbsnews.com/news/monsanto-pushes-golden-rice.

15. Ed Regis, "The True Story of the Genetically Modified Superfood That Almost Saved Millions," *Foreign Policy*, October 17, 2019, https://foreignpolicy.com/2019/10/17/golden-rice-genetically-modified-superfood-almost-saved-millions.

16. Robert Paarlberg, "A Dubious Success: The NGO Campaign Against GMOs," *GM Crops and Food* 5, no. 3 (November 6, 2014): 223–28, https://doi.org/10.4161/21645698.2014.952204.

17. Mark Lynas, "Anti-GMO Activists Lie About Attack on Rice Crop (and About So Many Other Things)," Slate, August 26, 2013, https://slate.com/technology/2013/08/golden-rice-attack-in-philippines-anti-gmo-activists-lie-about-protest-and-safety.html.

18. Regis, "The True Story of the Genetically Modified Superfood."

19. Joel Achenbach, "107 Nobel Laureates Sign Letter Blasting Greenpeace over GMOs," *Washington Post*, June 30, 2016, www.washingtonpost.com/news/speaking-of-science/wp/2016/06/29/more-than-100-nobel-laureates-take-on-greenpeace-over-gmo-stance.

20. Jessica Scarfuto, "Do You Trust Science? These Five Factors Play a Big Role," *Science*, February 16, 2020, www.sciencemag.org/news/2020/02/do-you-trust-science-these-five-factors-play-big-role.

21. Cary Funk, Alex Tyson, Brian Kennedy, and Courtney Johnson, "Scientists Are Among the Most Trusted Groups Internationally, Though Many Value Practical Experience over Expertise," Pew Research Center, September 29, 2020, www.pewresearch.org/science/2020/09/29/scientists-are-among-the-most-trusted-groups-in-society-though-many-value-practical-experience-over-expertise.

## 9 Exploring the Recently Plausible

1. Sam Meredith, "Brazil Braces for Renewed Covid Surge as Bolsonaro Faces Parliamentary Inquiry over Pandemic Response," CNBC, May 14, 2021, www.cnbc.com/2021/05/14/brazil-fears-third-covid-wave-as-bolsonaro-faces-parliamentary-inquiry.html.

2. Sanjeev Miglani and Devjyot Ghoshal, "PM Modi's Rating Falls to New Low as India Reels from COVID-19," Reuters, May 18, 2021, www.reuters.com/world/india/pm-modis-rating-falls-india-reels-covid-19-second-wave-2021-05-18.

3. "English Rendering of PM's Address at the World Economic Forum's Davos Dialogue," Press Information Bureau, Government of India, January 28, 2021, https://pib.gov.in/PressReleseDetail.aspx?PRID=1693019.

4. David Klepper and Neha Mehrotra, "Misinformation Surges amid India's COVID-19 Calamity," *Seattle Times*, May 13, 2021, www.seattletimes.com/business/misinformation-surges-amid-indias-covid-19-calamity.

## 10 Scenario One: Creating Your Child with Wellspring

1. Katsuhiko Hayashi, Orie Hikabe, Yayoi Obata, and Yuji Hirao, "Reconstitution of Mouse Oogenesis in a Dish from Pluripotent Stem Cells," *Nature Protocols* 12, no. 9 (September 2017): 1733–44, https://doi.org/10.1038/nprot.2017.070.

2. Tess Johnson, "Human Genetic Enhancement Might Soon Be Possible—but Where Do We Draw the Line?," The Conversation, December 3, 2019, http://theconversation.com/human-genetic-enhancement-might-soon-be-possible-but-where-do-we-draw-the-line-127406.

3. David Cyranoski, "The CRISPR-Baby Scandal: What's Next for Human Gene-Editing," *Nature* 566, no. 7745 (February 26, 2019): 440–42, https://doi.org/10.1038/d41586-019-00673-1.

4. Nathaniel Scharping, "How Are Neanderthals Different from Homo Sapiens?," *Discover*, May 5, 2020, www.discovermagazine.com/planet-earth/how-are-neanderthals-different-from-homo-sapiens.

5. Rachel Becker, "An Artificial Womb Successfully Grew Baby Sheep—and Humans Could Be Next," The Verge, April 25, 2017, www.theverge.com/2017/4/25/15421734/artificial-womb-fetus-biobag-uterus-lamb-sheep-birth-premie-preterm-infant.

6. Emily A. Partridge, Marcus G. Davey, Matthew A. Hornick, Patrick E. McGovern, Ali Y. Mejaddam, Jesse D. Vrecenak, Carmen Mesas-Burgos, et al., "An Extra-Uterine System to Physiologically Support the Extreme Premature Lamb," *Nature Communications* 8, no. 1 (April 25, 2017): 15112, https://doi.org/10.1038/ncomms15112.

7. Neera Bhatia and Evie Kendal, "We May One Day Grow Babies Outside the Womb, but There Are Many Things to Consider First," The Conversation, November 10, 2019, http://theconversation.com/we-may-one-day-grow-babies-outside-the-womb-but-there-are-many-things-to-consider-first-125709.

## 11 Scenario Two: What Happened When We Canceled Aging

1. "CRISPR/Cas9 Therapy Can Suppress Aging, Enhance Health and Extend Life Span in Mice," Science Daily, February 19, 2019, www.sciencedaily.com/releases/2019/02/190219111747.htm.

2. Chinese Academy of Sciences, "Scientists Develop New Gene Therapy Strategy to Delay Aging and Extend Lifespan," SciTechDaily, January 9, 2021, https://scitechdaily.com/scientists-develop-new-gene-therapy-strategy-to-delay-aging-and-extend-lifespan.

3. Adolfo Arranz, "Betting Big on Biotech," *South China Morning Post*, October 9, 2018, https://multimedia.scmp.com/news/china/article/2167415/china-2025-biotech/index.html.

4. Georgina M. Ellison-Hughes, "First Evidence That Senolytics Are Effective at Decreasing Senescent Cells in Humans," EBioMedicine, May 23, 2020, www.thelancet.com/journals/ebiom/article/PIIS2352-3964(19)30641-3/fulltext.

5. "CRISPR/Cas9 Therapy Can Suppress Aging."

6. Hughes, "First Evidence."

7. Amber Dance, "Science and Culture: The Art of Designing Life," *Proceedings of the National Academy of Sciences* 112, no. 49 (December 8, 2015): 14999–15001, https://doi.org/10.1073/pnas.1519838112.

8. Ning Zhang and Anthony A. Sauve, "Nicotinamide Adenine Dinucleotide," Science Direct, n.d., www.sciencedirect.com/topics/neuroscience/nicotinamide-adenine-dinucleotide.

9. Jared Friedman, "How Biotech Startup Funding Will Change in the Next 10 Years," YC Startup Library, n.d., www.ycombinator.com/library/4L-how-biotech-startup-funding-will-change-in-the-next-10-years.

10. Emily Mullin, "Five Ways to Get CRISPR into the Body," *MIT Technology Review*, September 22, 2017, www.technologyreview.com/2017/09/22/149011/five-ways-to-get-crispr-into-the-body.

11. We used historical S&P data and company financials from 2015 to 2020.

12. "Population Distribution by Age," Kaiser Family Foundation, 2019, www.kff.org/other/state-indicator/distribution-by-age/?currentTimeframe=0&sortModel=%7B%22colId%22:%22Location%22,%22sort%22:%22asc%22%7D.

13. "Policy Basics: The Supplemental Nutrition Assistance Program (SNAP)," Center on Budget and Policy Priorities, www.cbpp.org/research/food-assistance/the-supplemental-nutrition-assistance-program-snap.

14. "Trust Fund Data," Social Security, www.ssa.gov/oact/STATS/table4a3.html.

15. *Nijikai-jin* is a word Amy invented.

16. With apologies to Anthony Rizzo, who was arguably the Chicago Cubs' best first baseman of all time. Statistics from MLB.com.

17. "The Age Discrimination in Employment Act of 1967," US Equal Employment Opportunity Commission, www.eeoc.gov/statutes/age-discrimination-employment-act-1967.

### 12 Scenario Three: Akira Gold's "Where to Eat" 2037

1. Adam Platt, senior restaurant critic for *New York Magazine*, inspired this scenario. We imagined him in the year 2037, creating his annual "Where to Eat" guide.

2. Niina Heikkinen, "U.S. Bread Basket Shifts Thanks to Climate Change," *Scientific American*, December 23, 2015, www.scientificamerican.com/article/u-s-bread-basket-shifts-thanks-to-climate-change.

3. Euromonitor data, July 2020, www.euromonitor.com/usa.

4. "The Future of Agriculture: The Convergence of Tech and Bio Bringing Better Food to Market," SynBioBeta, February 9, 2020, https://synbiobeta.com/the-future-of-agriculture-the-convergence-of-tech-and-bio-bringing-better-food-to-market.

5. "Fermentation & Bioreactors," Sartorius, www.sartorius.com/en/products/fermentation-bioreactors.

6. Bioreactor market value data, Statista, February 2020, www.statista.com.

7. Gareth John Macdonald, "Bioreactor Design Adapts to Biopharma's Changing Needs," Genetic Engineering and Biotechnology News (GEN), July 1, 2019, www.genengnews.com/insights/bioreactor-design-adapts-to-biopharmas-changing-needs.

8. Senthold Asseng, Jose R. Guarin, Mahadev Raman, Oscar Monje, Gregory Kiss, Dickson D. Despommier, Forrest M. Meggers, and Paul P. G. Gauthier, "Wheat Yield Potential in Controlled-Environment Vertical Farms," *Proceedings of the National Academy of Sciences*, July 23, 2020, https://doi.org/10.1073/pnas.2002655117.

9. Karen Gilchrist, "This Multibillion-Dollar Company Is Selling Lab-Grown Chicken in a World-First," CNBC, March 1, 2021, www.cnbc.com/2021/03/01/eat-just-good-meat-sells-lab-grown-cultured-chicken-in-world-first.html.

10. Emily Waltz, "Club-Goers Take First Bites of Lab-Made Chicken," *Nature Biotechnology* 39, no. 3 (March 1, 2021): 257–58, https://doi.org/10.1038/s41587-021-00855-1.

11. Forecast for cultured meat by 2026, Source: BIS Research, April 2021.

12. Zoë Corbyn, "Out of the Lab and into Your Frying Pan: The Advance of Cultured Meat," *The Guardian*, January 19, 2020, www.theguardian.com/food/2020/jan/19/cultured-meat-on-its-way-to-a-table-near-you-cultivated-cells-farming-society-ethics.

13. Raito Ono, "Robotel: Japan Hotel Staffed by Robot Dinosaurs," Phys.org, August 31, 2018, https://phys.org/news/2018-08-robotel-japan-hotel-staffed-robot.html.

14. Global sales of service robots for professional use between 2018 and 2020. Source: IFR, September 2020.

15. James Borrell, "All Our Food Is 'Genetically Modified' in Some Way—Where Do You Draw the Line?," The Conversation, April 4, 2016, http://theconversation.com/all-our-food-is-genetically-modified-in-some-way-where-do-you-draw-the-line-56256.

16. Billy Lyons, "Is Molecular Whiskey the Futuristic Booze We've Been Waiting For?," *Fortune*, May 25, 2019, https://fortune.com/2019/05/25/endless-west-glyph-engineered-whiskey.

17. "Morpheus," DC Comics, February 29, 2012, www.dccomics.com/characters/morpheus.

18. Alice Liang, "World's First Molecular Whiskey Expands Its Portfolio," *Drinks Business*, November 5, 2020, www.thedrinksbusiness.com/2020/11/worlds-first-molecular-whiskey-expands-its-portfolio.

19. Nicole Trian, "Australia Prepares for 'Day Zero'—the Day the Water Runs Out," France 24, September 19, 2019, www.france24.com/en/20190919-australia-day-zero-drought-water-climate-change-greta-thunberg-paris-accord-extinction-rebe.

20. Kevin Winter, "Day Zero Is Meant to Cut Cape Town's Water Use: What Is It, and Is It Working?," The Conversation, February 20, 2018, http://theconversation.com/day-zero-is-meant-to-cut-cape-towns-water-use-what-is-it-and-is-it-working-92055.

21. Dave McIntyre, "It Was Only a Matter of Time. Lab-Created 'Molecular' Wine Is Here," *Washington Post*, March 6, 2020, www.washingtonpost.com/lifestyle/food/it-was-only-a-matter-of-time-lab-created-molecular-wine-is-here/2020/03/06/2f354ce8-5ef3-11ea-b014-4fafa866bb81_story.html.

22. Esther Mobley, "SF Startup Is Making Synthetic Wine in a Lab. Here's How It Tastes," *San Francisco Chronicle*, February 20, 2020, www.sfchronicle.com/wine/article/San-Francisco-startup-unveils-synthetic-wine-and-15068890.php.

23. Collin Dreizen, "Test-Tube Tasting? Bev Tech Company Unveils Grape-less 'Molecular Wine,'" *Wine Spectator*, February 26, 2020, www.winespectator.com/articles/test-tube-tasting-bev-tech-company-unveils-grape-less-molecular-wine-unfiltered.

**13 Scenario Four: The Underground**

1. The Underground was inspired by Coober Pedy, an Australian mining town where many people live in a subterranean community because the summers now top 120°F. Atlas Obscura offers a detailed overview of Coober Pedy at www.atlasobscura.com/places/coober-pedy. This scenario was also informed by *The Expanse* series by the writing duo James S. A. Corey and by Elon Musk's relentless desire to colonize Mars, which has been written about extensively.

2. "Climate Action Note—Data You Need to Know," United Nations Environment Programme, April 19, 2021, www.unep.org/explore-topics/climate-change/what-we-do/climate-action-note.

3. "The Paris Agreement," United Nations Framework Convention on Climate Change, https://unfccc.int/process-and-meetings/the-paris-agreement/the-paris-agreement.

4. "Transforming Food Systems," United Nations Environment Programme, April 20, 2021, www.unep.org/resources/factsheet/transforming-food-systems.

5. "Facts About the Climate Emergency," United Nations Environment Programme, January 25, 2021, www.unep.org/explore-topics/climate-change/facts-about-climate-emergency.

6. Mark Fischetti, "We Are Living in a Climate Emergency, and We're Going to Say So," *Scientific American*, April 12, 2021, www.scientificamerican.com/article/we-are-living-in-a-climate-emergency-and-were-going-to-say-so.

7. Mike Wall, "Elon Musk, X Prize Launch $100 Million Carbon-Removal Competition," Space.com, April 23, 2021, www.space.com/elon-musk-carbon-removal-x-prize.

8. Eric Berger, "Inside Elon Musk's Plan to Build One Starship a Week—and Settle Mars," Ars Technica, March 5, 2020, https://arstechnica.com/science/2020/03/inside-elon-musks-plan-to-build-one-starship-a-week-and-settle-mars.

9. Morgan McFall-Johnsen and Dave Mosher, "Elon Musk Says He Plans to Send 1 Million People to Mars by 2050 by Launching 3 Starship Rockets Every Day and Creating 'a Lot of Jobs' on the Red Planet," *Business Insider*, January 17, 2020, www.businessinsider.com/elon-musk-plans-1-million-people-to-mars-by-2050-2020-1.

10. Wall, "Elon Musk, X Prize Launch $100 Million Carbon-Removal Competition."

11. "Astronauts Answer Student Questions," NASA, www.nasa.gov/centers/johnson/pdf/569954main_astronaut%20_FAQ.pdf.

12. Eric Berger, "Meet the Real Ironman of Spaceflight: Valery Polyakov," Ars Technica, March 7, 2016, Valery Polyakov held the record for a single mission, spending an impressive 437 days on the Mir station in the 1990s.

13. "Longest Submarine Patrol," *Guinness Book of World Records*, www.guinnessworldrecords.com/world-records/submarine-patrol-longest.

14. Jackie Wattles, "Colonizing Mars Could Be Dangerous and Ridiculously Expensive. Elon Musk Wants to Do It Anyway," CNN, September 8, 2020, www.cnn.com/2020/09/08/tech/spacex-mars-profit-scn/index.html.

15. Gael Fashingbauer Cooper, "Elon Musk's First Name Shows Up in 1953 Book About Colonizing Mars," CNET, May 7, 2021, www.cnet.com/news/elon-musks-first-name-shows-up-in-1953-book-about-colonizing-mars.

16. Ali Bekhtaoui, "Egos Clash in Bezos and Musk Space Race," Phys.org, May 2, 2021, https://phys.org/news/2021-05-egos-clash-bezos-musk-space.html.

17. Sean O'Kane, "The Boring Company Tests Its 'Teslas in Tunnels' System in Las Vegas," The Verge, May 26, 2021, www.theverge.com/2021/5/26/22455365/elon-musk-boring-company-las-vegas-test-lvcc-loop-teslas.

18.Kathryn Hardison, "What Will Become of All This?," American City Business Journals, May 28, 2021, www.bizjournals.com/houston/news/2021/05/28/tesla-2500-acres-travis-county-plans.html.

19. Philip Ball, "Make Your Own World with Programmable Matter," IEEE Spectrum, May 27, 2014, https://spectrum.ieee.org/robotics/robotics-hardware/make-your-own-world-with-programmable-matter.

20. Neuralink website: https://neuralink.com.

21. Chia website: https://www.chia.net.

22. NOVOFARM website: https://www.f6s.com/novofarm.

23. Chris Impey, "This Is the Year the First Baby Will Be Born in Space," Inverse, May 30, 2021, www.inverse.com/science/when-will-the-first-baby-be-born-in-space.

24. Lisa Ruth Rand, "Colonizing Mars: Practicing Other Worlds on Earth," *Origins* 11, no. 2 (November 2017), https://origins.osu.edu/article/colonizing-mars-practicing-other-worlds-earth.

25. Derek Thompson, "Is Colonizing Mars the Most Important Project in Human History?," *The Atlantic*, June 29, 2018, www.theatlantic.com/technology/archive/2018/06/could-colonizing-mars-be-the-most-important-project-in-human-history/564041.

26. "What Is Biosphere 2," Biosphere 2, University of Arizona, https://biosphere2.org/visit/what-is-biosphere-2.

27. Our thinking about the EST economy and governing structure was loosely informed by Norway and Sweden. Interview with Dr. Christian Guilette, Scandinavian Faculty at University of California, Berkeley, April 23, 2021.

## 14 Scenario Five: The Memo

1. After reading a few papers, we were curious about which part of the US government would respond in the event of a cyber-bio attack. (The papers included Dor Farbiash and Rami Puzis, "Cyberbiosecurity: DNA Injection Attack in Synthetic Biology," ArXiv:2011.14224 [cs.CR], November 28, 2020, http://arxiv.org/abs/2011.14224; and Antonio Regalado, "Scientists Hack a Computer Using DNA," *MIT Technology Review*, August 10, 2017, www.technologyreview.com/2017/08/10/150013/scientists-hack-a-computer-using-dna.) We began by asking contacts at the US Department of Homeland Security and the Cybersecurity and Infrastructure Security Agency, discovering that neither organization had developed any protocol for such a situation. We pressed on, speaking with contacts in the US Air Force, US Navy, US Department of Defense, US State Department, US Government Accountably Office, and Centers for Disease Control and Prevention, as well as national security analysts and congressional staffers. A few contacts did walk us through the step-by-step process that would be involved in the event of a cyber-bio attack. Here was a representative response, and it's worth noting that no two people gave us the same answers:

"It's a great question. I suspect this would look quite similar to COVID's response, in that an interagency task force would be created at the National Security Council. Your top officials there would likely be the 4-star general who dual-hats as US Cyber Command's commander and runs the National Security Agency. That's for the cyber element. The CDC, HHS, and NIH would be brought in, the State Department to see what we knew about the Chinese lab that did the [work], and ultimately the FBI to conduct any necessary domestic investigations (assume this left-wing group is US based). The US national security advisor, or more likely the Deputy NSA, would lead the task force. If it were serious enough, the Vice President would take over leadership of the task force."

## 15 A New Beginning

1. "Park History," Asilomar Conference Grounds, www.visitasilomar.com/discover/park-history.

2. Paul Berg, David Baltimore, Herbert W. Boyer, Stanley N. Cohen, Ronald W. Davis, David S. Hogness, Daniel Nathans, et al., "Potential Biohazards of Recombinant DNA Molecules," *Science* 185, no. 4148 (July 26, 1974): 303, https://doi.org/10.1126/science.185.4148.303.

3. Nicolas Rasmussen, "DNA Technology: 'Moratorium' on Use and Asilomar Conference," Wiley Online Library, January 27, 2015, https://onlinelibrary.wiley.com/doi/abs/10.1002/9780470015902.a0005613.pub2.

4. "Transcript of Nixon's Address on Troop Withdrawals and Situation in Vietnam," *New York Times*, April 27, 1972, www.nytimes.com/1972/04/27/archives/transcript-of-nixons-address-on-troop-withdrawals-and-situation-in.html.

5. Douglas MacEachin, "Predicting the Soviet Invasion of Afghanistan: The Intelligence Community's Record," Center for the Study of Intelligence Monograph, March 2003, posted at Federation of American Scientists, Intelligence Resource Program, https://fas.org/irp/cia/product/afghanistan/index.html.

6. "A Guide to the United States' History of Recognition, Diplomatic, and Consular Relations, by Country, Since 1775: China," US Department of State, Office of the Historian, https://history.state.gov/countries/china/china-us-relations.

7. Ashley M. Eskew and Emily S. Jungheim, "A History of Developments to Improve in Vitro Fertilization," *Missouri Medicine* 114, no. 3 (2017): 156–59, full text at National Center for Biotechnology Information, www.ncbi.nlm.nih.gov/pmc/articles/PMC6140213.

8. Ariana Eunjung Cha, "40 Years After 1st 'Test Tube' Baby, Science Has Produced 7 Million Babies—and Raised Moral Questions," *Chicago Tribune*, April 27, 2018, www.chicagotribune.com/lifestyles/parenting/ct-test-tube-babies-moral-questions-20180427-story.html.

9. Institute of Medicine (US) Committee to Study Decision Making; Hanna KE, editor, "Asilomar and Recombinant DNA: The End of the Beginning," *Biomedical Politics*, Washington (DC): National Academies Press (US), 1991, www.ncbi.nlm.nih.gov/books/NBK234217.

10. Institute of Medicine, *Biomedical Politics.*

11. Institute of Medicine, *Biomedical Politics.*

12. Institute of Medicine, *Biomedical Politics.*

13. Michael Rogers, "The Pandora's Box Congress," *Rolling Stone*, June 19, 1975, 37–42, 74–82.

14. Dan Ferber, "Time for a Synthetic Biology Asilomar?," *Science* 303, no. 5655 (January 9, 2004): 159, https://doi.org/10.1126/science.303.5655.159.

15. Richard Harris, "The Presidency and the Press," *New Yorker*, September 24, 1973, www.newyorker.com/magazine/1973/10/01/the-presidency-and-the-press.

16. "Edelman Trust Barometer 2021," Edelman, www.edelman.com/trust/2021-trust-barometer.

17. Tomi Kilgore, "Ginkgo Bioworks to Be Taken Public by SPAC Soaring Eagle at a Valuation of $15 Billion," MarketWatch, May 11, 2021, www.marketwatch com/story/ginkgo-bioworks-to-be-taken-public-by-spac-soaring-eagle-at-a-valuation-of-15-billion-2021-05-11.

18. "New Jersey Coronavirus Update: Rutgers Students Protest COVID-19 Vaccine Requirement," ABC7 New York, May 21, 2021, https://abc7ny.com/health/rutgers-students-protest-covid-19-vaccine-requirement-/10672983.

19. Brad Smith, "The Need for a Digital Geneva Convention," Microsoft, February 14, 2017, https://blogs.microsoft.com/on-the-issues/2017/02/14/need-digital-geneva-convention.

20. Romesh Ratnesar, "How Microsoft's Brad Smith is Trying to Restore Your Trust in Big Tech," Time.com, September 9, 2019, https://time.com/5669537/brad-smith-microsoft-big-tech.

21. Bill Gates, "Here's My Plan to Improve Our World—and How You Can Help," *Wired*, November 12, 2013, www.wired.com/2013/11/bill-gates-wired-essay.

22. "News, Trends, and Stories from the Synthetic Biology Industry," Synbiobeta Digest, August 2019, https://synbiobeta.com/wp-content/uploads/2019/08/Digest-288.html.

23. "Broad Institute Launches the Eric and Wendy Schmidt Center to Connect Biology, Machine Learning for Understanding Programs of Life," Broad Institute, March 25, 2021, www.broadinstitute.org/news/broad-institute-launches-eric-and-wendy-schmidt-center-connect-biology-machine-learning.

24. "China Focus: China Stepping Closer to 'Innovative Nation,'" Xinhua, May 5, 2017, www.xinhuanet.com/english/2017-05/05/c_136260598.htm.

25. Simon Johnson, "China, the Innovation Dragon," Peterson Institute for International Economics, January 3, 2018, www.piie.com/blogs/china-economic-watch/china-innovation-dragon.

26. Ayala Ochert, "National Gene Bank Opens in China," BioNews, September 26, 2016, www.bionews.org.uk/page_95701.

27. See, for example, a sample of search results from ClinicalTrials.gov, US National Library of Medicine, https://clinicaltrials.gov/ct2/results?cond=cancer+&term=crispr&cntry=CN&state=&city=&dist=.

28. Elsa B. Kania and Wilson Vorndick, "Weaponizing Biotech: How China's Military Is Preparing for a 'New Domain of Warfare,'" Defense One, August 14, 2019, www.defenseone.com/ideas/2019/08/chinas-military-pursuing-biotech/159167.

29. "Yuan Longping Died on May 22nd," *The Economist*, May 29, 2021, www.economist.com/obituary/2021/05/29/yuan-longping-died-on-may-22nd.

30. Keith Bradsher and Chris Buckley, "Yuan Longping, Plant Scientist Who Helped Curb Famine, Dies at 90," *New York Times*, May 23, 2021, www.nytimes.com/2021/05/23/world/asia/yuan-longping-dead.html.

31. Li Yuan and Rumsey Taylor, "How Thousands in China Gently Mourn a Coronavirus Whistle-Blower," *New York Times*, April 13, 2020, www.nytimes.com/interactive/2020/04/13/technology/coronavirus-doctor-whistleblower-weibo.html.

32. Shannon Ellis, "Biotech Booms in China," *Nature* 553, no. 7688 (January 17, 2018): S19–22, https://doi.org/10.1038/d41586-018-00542-3.

33. James McBride and Andrew Chatzky, "Is 'Made in China 2025' a Threat to Global Trade?," Council on Foreign Relations, updated May 13, 2019, www.cfr.org/backgrounder/made-china-2025-threat-global-trade.

34. "The World in 2050," PricewaterhouseCoopers, www.pwc.com/gx/en/research-insights/economy/the-world-in-2050.html.

35. Renu Swarup, "Biotech Nation: Support for Innovators Heralds a New India," Nature India, April 30, 2018, www.natureasia.com/en/nindia/article/10.1038/nindia.2018.55.

36. Meredith Wadman, "Falsified Data Gets India's Largest Generic Drug-Maker into Trouble," *Nature*, March 2, 2009, https://doi.org/10.1038/news.2009.130.

37. "New Israeli Innovation Box Regime: An Update and Review of Key Features," Ernst and Young, Tax News Update, May 31, 2019, https://taxnews.ey.com/news/2019-1022-new-israeli-innovation-box-regime-an-update-and-review-of-key-features.

38. Endless Possibilities to Promote Innovation brochure, available as a PDF from https://innovationisrael.org.il.

39. Aradhana Aravindan and John Geddie, "Singapore Approves Sale of Lab-Grown Meat in World First," Reuters, December 2, 2020, www.reuters.com/article/us-eat-just-singapore-idUKKBN28C06Z.

40. Patrice Laget and Mark Cantley, "European Responses to Biotechnology: Research, Regulation, and Dialogue," *Issues in Science and Technology* 17, no. 4 (Summer 2001), https://issues.org/laget.

41. Jenny Howard, "Plague Was One of History's Deadliest Diseases—Then We Found a Cure," *National Geographic*, July 6, 2020, www.nationalgeographic.com/science/article/the-plague.

42. Nidhi Subbaraman, "US Officials Revisit Rules for Disclosing Risky Disease Experiments," *Nature*, January 27, 2020, https://doi.org/10.1038/d41586-020-00210-5.

43. Sandra Kollen Ghizoni, "Creation of the Bretton Woods System," Federal Reserve History, November 22, 2013, www.federalreservehistory.org/essays/bretton-woods-created.

44. Michael Bordo, Owen Humpage, and Anna J. Schwartz, "U.S. Intervention During the Bretton Wood Era, 1962–1973," Working Paper 11-08, Federal Reserve Bank of Cleveland, www.clevelandfed.org/en/newsroom-and-events/publications/working-papers/2011-working-papers/wp-1108-us-intervention-during-the-bretton-woods-era-1962-to-1973.aspx.

45. "DNA," Interpol, www.interpol.int/en/How-we-work/Forensics/DNA.

46. "Population, Total—Estonia," World Bank, https://data.worldbank.org/indicator/SP.POP.TOTL?locations=EE.

47. "Estonia," Place Explorer, Data Commons, https://datacommons.org/place/country/EST?utm_medium=explore&mprop=count&popt=Person&hl.

48. "The Estonian Biobank," EIT Health Scandinavia, www.eithealth-scandinavia.eu/biobanks/the-estonian-biobank.

49. "International Driving Permit," AAA, www.aaa.com/vacation/idpf.html.

50. George M. Church and Edward Regis, *Regenesis: How Synthetic Biology Will Reinvent Nature and Ourselves* (New York: Basic Books, 2014).

51. "FBI Laboratory Positions," Federal Bureau of Investigation, www.fbi.gov/services/laboratory/laboratory-positions.

52. "New Cyberattack Can Trick Scientists into Making Dangerous Toxins or Synthetic Viruses, According to BGU Cyber-Researchers," Ben-Gurion University of the Negev, November 30, 2020, https://in.bgu.ac.il/en/pages/news/toxic_viruses.aspx.

53. Rami Puzis, Dor Farbiash, Oleg Brodt, Yuval Elovici, and Dov Greenbaum, "Increased Cyber-Biosecurity for DNA Synthesis," *Nature Biotechnology* 38, no. 12 (December 2020): 1379–81, https://doi.org/10.1038/s41587-020-00761-y.

54. Islamorada, Florida, town council website: https://www.islamorada.fl.us/village_council/index.php.

55. Amy Webb interviewed John Cumbers on May 20, 2021.

56. Megan Molteni, "23andMe's Pharma Deals Have Been the Plan All Along," *Wired*, August 3, 2018, www.wired.com/story/23andme-glaxosmithkline-pharma-deal.

57. Ben Stevens, "Waitrose Launches DNA Test Pop-Ups Offering Shoppers Personal Genetic Health Advice," Charged, December 3, 2019, www.chargedretail.co.uk/2019/12/03/waitrose-launches-dna-test-pop-ups-offering-shoppers-personal-genetic-health-advice.

58. Catherine Lamb, "CES 2020: DNANudge Guides Your Grocery Shopping Based Off of Your DNA," The Spoon, January 7, 2020, https://thespoon.tech/dnanudge-guides-your-grocery-shopping-based-off-of-your-dna.

59. Brian Knutson, Scott Rick, G. Elliott Wimmer, Drazen Prelec, and George Loewenstein, "Neural Predictors of Purchases," *Neuron* 53, no. 1 (January 4, 2007): 147–56, https://doi.org/10.1016/j.neuron.2006.11.010.

60. "Researchers Use Brain Scans to Predict When People Will Buy Products," Carnegie Mellon University, January 3, 2007, press release, posted at EurekAlert, American Association for the Advancement of Science, www.eurekalert.org/pub_releases/2007-01/cmu-rub010307.php.

61. Carl Williott, "What's Better, Sex or Shopping? Your Brain Doesn't Know and Doesn't Care," MTV News, www.mtv.com/news/2134197/shopping-sex-brain-study.

62. "FAQs About 'Resource Profile and User Guide of the Polygenic Index Repository,'" Social Science Genetic Association Consortium, www.thessgac.org/faqs.

63. Nanibaa' A. Garrison, "Genomic Justice for Native Americans: Impact of the Havasupai Case on Genetic Research," *Science, Technology and Human Values* 38, no. 2 (2013): 201–23, https://doi.org/10.1177/0162243912470009.

64. Amy Harmon, "Indian Tribe Wins Fight to Limit Research of Its DNA," *New York Times*, April 21, 2010, www.nytimes.com/2010/04/22/us/22dna.html.

65. Sara Reardon, "Navajo Nation Reconsiders Ban on Genetic Research," *Nature* 550, no. 7675 (October 6, 2017): 165–66, www.nature.com/news/navajo-nation-reconsiders-ban-on-genetic-research-1.22780.

66. "The Legacy of Henrietta Lacks," Johns Hopkins Medicine, www.hopkinsmedicine.org/henriettalacks.

67. "The Tuskegee Timeline," The U.S. Public Health Service Syphilis Study at Tuskegee, CDC.com, www.cdc.gov/tuskegee/timeline.htm.

68. "Need to Increase Diversity Within Genetic Data Sets: Diversifying Population-Level Genetic Data Beyond Europeans Will Expand the Power of Polygenic Scores," Science Daily, March 29, 2019, www.sciencedaily.com/releases/2019/03/190329134743.htm.

69. Data from the All of Us Research Program, National Institutes of Health, https://allofus.nih.gov.

70. Katherine J. Wu, "Scientific Journals Commit to Diversity but Lack the Data," *New York Times*, October 30, 2020, www.nytimes.com/2020/10/30/science/diversity-science-journals.html.

71. "Staff and Advisory Board," *Cell*, www.cell.com/cell/editorial-board, accessed May 15, 2021.

# *BIBLIOGRAPHY*

THIS IS AN ABRIDGED BIBLIOGRAPHY. TO VIEW THE COMPLETE LIST OF sources used during our research and writing, visit our repository of further information at Dropbox (http://bit.ly/GenesisMachine) or scan the QR code below.

Abbott, Timothy R., Girija Dhamdhere, Yanxia Liu, Xueqiu Lin, Laine Goudy, Leiping Zeng, Augustine Chemparathy, et al. "Development of CRISPR as an Antiviral Strategy to Combat SARS-CoV-2 and Influenza." *Cell* 181, no. 4 (May 14, 2020): 865–76.e12. https://doi.org/10.1016/j.cell.2020.04.020.

"About the Protocol." Convention on Biological Diversity, https://bch.cbd.int/protocol/background.

Agius, E. "Germ-Line Cells: Our Responsibilities for Future Generations." In *Our Responsibilities Towards Future Generations*, ed. S. Busuttil. Malta: Foundation for International Studies, 1990.

Ahammad, Ishtiaque, and Samia Sultana Lira. "Designing a Novel mRNA Vaccine Against SARS-CoV-2: An Immunoinformatics Approach." *International Journal of Biological Macromolecules* 162 (November 1, 2020): 820–37. https://doi.org/10.1016/j.ijbiomac.2020.06.213.

Akbari, Omar S., Hugo J. Bellen, Ethan Bier, Simon L. Bullock, Austin Burt, George M. Church, Kevin R. Cook, et al. "Safeguarding Gene Drive Experiments in the Laboratory." *Science* 349 (2015): 972–79.

Alem, Sylvain, Clint J. Perry, Xingfu Zhu, Olli J. Loukola, Thomas Ingraham, Eirik Sovik, and Lars Chittka. "Associative Mechanisms Allow for Social Learning and Cultural Transmission of String Pulling in an Insect." *PLOS Biology* 14 (2016): el00256.

Alivisatos, A. Paul, Miyoung Chun, George M. Church, Ralph J. Greenspan, Michael L. Roukes, and Rafael Yuste. "The Brain Activity Map Project and the Challenge of Functional Connectomics." *Neuron* 74, no. 6 (June 21, 2012): 970–74. https://doi.org/10.1016/j.neuron.2012.06.006.

———. "A National Network of Neurotechnology Centers for the BRAIN Initiative." *Neuron* 88, no. 3 (2015): 445–48. https://doi.org/10.1016/j.neuron.2015.10.015.

Allen, Garland. "Eugenics and Modern Biology: Critiques of Eugenics, 1910–1945." *Annals of Human Genetics* 75 (2011): 314–25.

———. "Mendel and Modern Genetics: The Legacy for Today." *Endeavour* 27 (2003): 63–68.

Andersen, Ross. "Welcome to Pleistocene Park." *The Atlantic*, April 2017, www.theatlantic.com/magazine/archive/2017/04/pleistocene-park/517779.

Anderson, Sam. "The Last Two Northern White Rhinos on Earth." *New York Times*, January 6, 2021, www.nytimes.com/2021/01/06/magazine/the-last-two-northern-white-rhinos-on-earth.html.

Andrianantoandro, Ernesto. "Manifesting Synthetic Biology." *Trends in Biotechnology* 33, no. 2 (February 1, 2015): 55–56. https://doi.org/10.1016/j.tibtech.2014.12.002.

Arkin, Adam. "Setting the Standard in Synthetic Biology." *Nature Biotechnology* 26, no. 7 (July 2008): 771–74. https://doi.org/10.1038/nbt0708-771.

Asseng, Senthold, Jose R. Guarin, Mahadev Raman, Oscar Monje, Gregory Kiss, Dickson D. Despommier, Forrest M. Meggers, and Paul P. G. Gauthier. "Wheat Yield Potential in Controlled-Environment Vertical Farms." *Proceedings of the National Academy of Sciences*, July 23, 2020. https://doi.org/10.1073/pnas.2002655117.

Ball, Philip. "The Patent Threat to Designer Biology." *Nature*, June 22, 2007. https://doi.org/10.1038/news070618-17.

Baltes, Nicholas J., and Daniel F. Voytas. "Enabling Plant Synthetic Biology Through Genome Engineering." *Trends in Biotechnology* 33, no. 2 (February 1, 2015): 120–31. https://doi.org/10.1016/j.tibtech.2014.11.008.

Bartley, Bryan, Jacob Beal, Kevin Clancy, Goksel Misirli, Nicholas Roehner, Ernst Oberortner, Matthew Pocock, et al. "Synthetic Biology Open Language (SBOL)

Version 2.0.0." *Journal of Integrative Bioinformatics* 12, no. 2 (June 1, 2015): 902–91. https://doi.org/10.1515/jib-2015-272.

Bartley, Bryan A., Jacob Beal, Jonathan R. Karr, and Elizabeth A. Strychalski. "Organizing Genome Engineering for the Gigabase Scale." *Nature Communications* 11, no. 1 (February 4, 2020): 689. https://doi.org/10.1038/s41467-020-14314-z.

Beal, Jacob, Traci Haddock-Angelli, Natalie Farny, and Randy Rettberg. "Time to Get Serious About Measurement in Synthetic Biology." *Trends in Biotechnology* 36, no. 9 (September 1, 2018): 869–71. https://doi.org/10.1016/j.tibtech.2018.05.003.

Belluck, Pam. "Chinese Scientist Who Says He Edited Babies' Genes Defends His Work." *New York Times*, November 28, 2018, www.nytimes.com/2018/11/28/world/asia/gene-editing-babies-he-jiankui.html.

Benner, Steven A. "Synthetic Biology: Act Natural." *Nature* 421, no. 6919 (January 2003): 118. https://doi.org/10.1038/421118a.

Berg, Paul, David Baltimore, Herbert W. Boyer, Stanley N. Cohen, Ronald W. Davis, David S. Hogness, Daniel Nathans, et al. "Potential Biohazards of Recombinant DNA Molecules." *Science* 185, no. 4148 (July 26, 1974): 303. https://doi.org/10.1126/science.185.4148.303.

Bettinger, Blaine. "Esther Dyson and the 'First 10.'" The Genetic Genealogist, July 27, 2007, https://thegeneticgenealogist.com/2007/07/27/esther-dyson-and-the-first-10.

Bhattacharya, Shaoni. "Stupidity Should Be Cured, Says DNA Discoverer." *New Scientist*, February 28, 2003, www.newscientist.com/article/dn3451-stupidity-should-be-cured-says-dna-discoverer.

Biello, David. "3 Billion to Zero: What Happened to the Passenger Pigeon?" *Scientific American*, June 27, 2014, www.scientificamerican.com/article/3-billion-to-zero-what-happened-to-the-passenger-pigeon.

Billiau, Alfons. "At the Centennial of the Bacteriophage: Reviving the Overlooked Contribution of a Forgotten Pioneer, Richard Bruynoghe (1881–1957)." *Journal of the History of Biology* 49, no. 3 (August 1, 2016): 559–80. https://doi.org/10.1007/s10739-015-9429-0.

"Biosecurity and Dual-Use Research in the Life Sciences," in National Research Council, Committee on a New Government-University Partnership for Science and Security, *Science and Security in a Post 9/11 World: A Report Based on Regional Discussions Between the Science and Security Communities*. Washington, DC: National Academies Press, 2007, 57–68, www.ncbi.nlm.nih.gov/books/NBK11496.

Birch, Douglas. "Race for the Genome." *Baltimore Sun*, May 18, 1999.

Blake, William J., and Farren J. Isaacs. "Synthetic Biology Evolves." *Trends in Biotechnology* 22, no. 7 (July 1, 2004): 321–24. https://doi.org/10.1016/j.tibtech.2004.04.008.

Blendon, Robert J., Mary T. Gorski, and John M. Benson. "The Public and the Gene-Editing Revolution." *New England Journal of Medicine* 374, no. 15 (April 14, 2016): 1406–11. https://doi.org/10.1056/NEJMp1602010.

Bonnet, Jérôme, and Drew Endy. "Switches, Switches, Every Where, in Any Drop We Drink." *Molecular Cell* 49, no. 2 (January 24, 2013): 232–33. https://doi.org/10.1016/j.molcel.2013.01.005.

Borrell, James. "All Our Food Is 'Genetically Modified' in Some Way—Where Do You Draw the Line?" The Conversation, April 4, 2016, http://theconversation.com/all-our-food-is-genetically-modified-in-some-way-where-do-you-draw-the-line-56256.

Brandt, K., and R. Barrangou. "Applications of CRISPR Technologies Across the Food Supply Chain." *Annual Review of Food Sciences Technology* 10, no. 133 (2019).

Bueno de Mesquita, B., and A. Smith. *The Dictator's Handbook: Why Bad Behavior Is Almost Always Good Politics*. New York: PublicAffairs, 2012.

Bueso, Yensi Flores, and Mark Tangney. "Synthetic Biology in the Driving Seat of the Bioeconomy." *Trends in Biotechnology* 35, no. 5 (May 1, 2017): 373–78. https://doi.org/10.1016/j.tibtech.2017.02.002.

Büllesbach, Erika E., and Christian Schwabe. "The Chemical Synthesis of Rat Relaxin and the Unexpectedly High Potency of the Synthetic Hormone in the Mouse." *European Journal of Biochemistry* 241, no. 2 (1996): 533–37. https://doi.org/10.1111/j.1432-1033.1996.00533.x.

Burkhardt, Peter K., Peter Beyer, Joachim Wünn, Andreas Klöti, Gregory A. Armstrong, Michael Schledz, Johannes von Lintig, and Ingo Potrykus. "Transgenic Rice (*Oryza sativa*) Endosperm Expressing Daffodil (*Narcissus pseudonarcissus*) Phytoene Synthase Accumulates Phytoene, a Key Intermediate of Provitamin A Biosynthesis." *Plant Journal* 11, no. 5 (1997): 1071–78. https://doi.org/10.1046/j.1365-313X.1997.11051071.x.

Caliendo, Angela M., and Richard L. Hodinka. "A CRISPR Way to Diagnose Infectious Diseases." *New England Journal of Medicine* 377, no. 17 (October 26, 2017): 1685–87. https://doi.org/10.1056/NEJMcibr1704902.

Callaway, Ewen. "Small Group Scoops International Effort to Sequence Huge Wheat Genome." *Nature News*, October 31, 2017. https://doi.org/10.1038/nature.2017.22924.

Calos, Michele P. "The CRISPR Way to Think About Duchenne's." *New England Journal of Medicine* 374, no. 17 (April 28, 2016): 1684–86. https://doi.org/10.1056/NEJMcibr1601383.z

Carlson, Robert H. *Biology Is Technology: The Promise, Peril, and New Business of Engineering Life*. Cambridge, MA: Harvard University Press, 2010.

Carrington, Damian. "Giraffes Facing Extinction After Devastating Decline, Experts Warn." *The Guardian*, December 8, 2016, www.theguardian.com/environment/2016/dec/08/giraffe-red-list-vulnerable-species-extinction.

Carter, William. Statement Before the House Armed Services Committee, Subcommittee on Emerging Threats and Capabilities, 115th Cong., 2nd sess., January 9, 2018, Homeland Security Digital Library, www.hsdl.org/?abstract&did=822422.

Ceballos, Gerardo, Paul R. Ehrlich, Anthony D. Barnosky, Andrés García, Robert M. Pringle, and Todd M. Palmer. "Accelerated Modern Human–Induced

Species Losses: Entering the Sixth Mass Extinction." *Science Advances* 1, no. 5 (June 2015): e1400253. https://doi.org/10.1126/sciadv.1400253.

"Celera Wins Genome Race." *Wired*, April 6, 2000, www.wired.com/2000/04/celera-wins-genome-race.

Cha, Ariana Eunjung. "Companies Rush to Build 'Biofactories' for Medicines, Flavorings and Fuels." *Washington Post*, October 24, 2013, www.washingtonpost.com/national/health-science/companies-rush-to-build-biofactories-for-medicines-flavorings-and-fuels/2013/10/24/f439dc3a-3032-11e3-8906-3daa2bcde110_story.html.

Chadwick, B. P., L. J. Campbell, C. L. Jackson, L. Ozelius, S. A. Slaugenhaupt, D. A. Stephenson, J. H. Edwards, J. Wiest, and S. Povey. "Report on the Sixth International Workshop on Chromosome 9 Held at Denver, Colorado, 27 October 1998." *Annals of Human Genetics* 63, no. 2 (1999): 101–17. https://doi.org/10.1046/j.1469-1809.1999.6320101.x.

Chalmers, D. J. *The Conscious Mind: In Search of a Fundamental Theory*. Philosophy of Mind Series. New York: Oxford University Press, 1996.

Check, Erika. "Synthetic Biologists Try to Calm Fears." *Nature* 441, no. 7092 (May 1, 2006): 388–89. https://doi.org/10.1038/441388a.

Chen, Ming, and Dan Luo. "A CRISPR Path to Cutting-Edge Materials." *New England Journal of Medicine* 382, no. 1 (January 2, 2020): 85–88. https://doi.org/10.1056/NEJMcibr1911506.

Chen, Shi-Lin, Hua Yu, Hong-Mei Luo, Qiong Wu, Chun-Fang Li, and André Steinmetz. "Conservation and Sustainable Use of Medicinal Plants: Problems, Progress, and Prospects." *Chinese Medicine* 11 (July 30, 2016). https://doi.org/10.1186/s13020-016-0108-7.

Chien, Wade W. "A CRISPR Way to Restore Hearing." *New England Journal of Medicine* 378, no. 13 (March 29, 2018): 1255–56. https://doi.org/10.1056/NEJMcibr1716789.

Cho, Renee. "How Climate Change Will Alter Our Food." State of the Planet, Columbia Climate School, July 25, 2018, https://blogs.ei.columbia.edu/2018/07/25/climate-change-food-agriculture.

Christiansen, Jen. "Gene Regulation, Illustrated." Scientific American Blog Network, May 12, 2016, https://blogs.scientificamerican.com/sa-visual/gene-regulation-illustrated.

Christensen, Jon. "Scientist at Work. Ingo Potrykus: Golden Rice in a Grenade-Proof Greenhouse." *New York Times*, November 21, 2000, www.nytimes.com/2000/11/21/science/scientist-at-work-ingo-potrykus-golden-rice-in-a-grenade-proof-greenhouse.html.

Church, George. "Compelling Reasons for Repairing Human Germlines." *New England Journal of Medicine* 377, no. 20 (November 16, 2017): 1909–11. https://doi.org/10.1056/NEJMp1710370.

———. "Genomes for All." *Scientific American*, January 2006, www.scientificamerican.com/article/genomes-for-all. https://doi.org/10.1038/scientificamerican0106-46.

———. "George Church: De-Extinction Is a Good Idea." *Scientific American*, September 1, 2013, www.scientificamerican.com/article/george-church-de-extinction-is-a-good-idea. https://doi.org/10.1038/scientificamerican0913-12.

Church, George, and Ed Regis. *Regenesis: How Synthetic Biology Will Reinvent Nature and Ourselves*. New York: Basic Books, 2014.

Clarke, Arthur C. "Extra-Terrestrial Relays: Can Rocket Stations Give World-Wide Radio Coverage?" In *Progress in Astronautics and Rocketry*, ed. Richard B. Marsten, 19: 3–6. Communication Satellite Systems Technology. Amsterdam: Elsevier, 1966. https://doi.org/10.1016/B978-1-4832-2716-0.50006-2.

"Cloning Insulin." Genentech, April 7, 2016, www.gene.com/stories/cloning-insulin.

Coffey, Rebecca. "Bison versus Mammoths: New Culprit in the Disappearance of North America's Giants." *Scientific American*, www.scientificamerican.com/article/bison-vs-mammoths.

Cohen, Jacques, and Henry Malter. "The First Clinical Nuclear Transplantation in China: New Information About a Case Reported to ASRM in 2003." Reproductive BioMedicine Online 33, no. 4 (October 1, 2016): 433–35. https://doi.org/10.1016/j.rbmo.2016.08.002.

Cohen S. N., A. C. Chang, H. W. Boyer, and R. B. Helling. "Construction of Biologically Functional Bacterial Plasmids *in Vitro*." *Proceedings of the National Academy of Sciences* 70, no. 11 (November 1, 1973): 3240–44. https://doi.org/10.1073/pnas.70.11.3240.

Coley, Conner W., Dale A. Thomas III, Justin A.M. Lummiss, Jonathan N. Jaworski, Christopher P. Breen, Victor Schultz, Travis Hart, et al. "A Robotic Platform for Flow Synthesis of Organic Compounds Informed by AI Planning." *Science* 365, no. 6453 (August 2019).

Committee on Strategies for Identifying and Addressing Potential Biodefense Vulnerabilities Posed by Synthetic Biology, Board on Chemical Sciences and Technology, Board on Life Sciences, Division on Earth and Life Studies, and National Academies of Sciences, Engineering, and Medicine. *Biodefense in the Age of Synthetic Biology*. Washington, DC: National Academies Press, 2018. https://doi.org/10.17226/24890.

Coxworth, Ben. "First Truly Synthetic Organism Created Using Four Bottles of Chemicals and a Computer." *New Atlas*, May 21, 2010, https://newatlas.com/first-synthetic-organism-created/15165.

Cravens, A., J. Payne, and C. D. Smolke. "Synthetic Biology Strategies for Microbial Biosynthesis of Plant Natural Products." *Nature Communications* 10, no. 2142 (May 13, 2019).

Cyranoski, David. "What CRISPR-Baby Prison Sentences Mean for Research." *Nature* 577, no. 7789 (January 3, 2020): 154–55. https://doi.org/10.1038/d41586-020-00001-y.

Dance, Amber. "Science and Culture: The Art of Designing Life." *Proceedings of the National Academy of Sciences* 112, no. 49 (December 8, 2015): 14999–15001. https://doi.org/10.1073/pnas.1519838112.

Davey, Melissa. "Scientists Sequence Wheat Genome in Breakthrough Once Thought 'Impossible.'" *The Guardian*, August 16, 2018, www.theguardian

.com/science/2018/aug/16/scientists-sequence-wheat-genome-in-breakthrough-once-thought-impossible.

Diamond, Jared. *Collapse: How Societies Choose to Fail or Succeed*, rev. ed. New York: Penguin, 2011.

Dolgin, Elie. "Synthetic Biology Speeds Vaccine Development." *Nature Research*, September 28, 2020. https://doi.org/10.1038/d42859-020-00025-4.

Doudna, Jennifer A., and Samuel H. Sternberg. *A Crack in Creation: Gene Editing and the Unthinkable Power to Control Evolution*. Boston: Houghton Mifflin Harcourt, 2017.

Dowdy, Steven F. "Controlling CRISPR-Cas9 Gene Editing." *New England Journal of Medicine* 381, no. 3 (July 18, 2019): 289–90. https://doi.org/10.1056/NEJMcibr1906886.

Drexler, Eric K. *Engines of Creation—The Coming Era of Nanotechnology*. New York: Anchor, 1987.

Duhaime-Ross, Arielle. "In Search of a Healthy Gut, One Man Turned to an Extreme DIY Fecal Transplant." The Verge, May 4, 2016, www.theverge.com/2016/5/4/11581994/fmt-fecal-matter-transplant-josiah-zayner-microbiome-ibs-c-diff.

Dyson, Esther. "Full Disclosure." *Wall Street Journal*, July 25, 2007, www.wsj.com/articles/SB118532736853177075.

Dyson, George B. *Darwin Among the Machines: The Evolution of Global Intelligence*. New York: Basic Books, 1997.

Eden, A., J. Søraker, J. H. Moor, and E. Steinhart, eds. *Singularity Hypotheses: A Scientific and Philosophical Assessment*. The Frontiers Collection. Berlin: Springer, 2012.

Editors, The. "Why Efforts to Bring Extinct Species Back from the Dead Miss the Point." *Scientific American*, June 1, 2013, www.scientificamerican.com/article/why-efforts-bring-extinct-species-back-from-dead-miss-point.

Ellison-Hughes, Georgina M. "First Evidence That Senolytics Are Effective at Decreasing Senescent Cells in Humans." EBioMedicine, May 23, 2020, www.thelancet.com/journals/ebiom/article/PIIS2352-3964(19)30641-3/fulltext.

Endy, Drew. "Foundations for Engineering Biology." *Nature* 438, no. 7067 (November 2005): 449–53. https://doi.org/10.1038/nature04342.

"Engineered Swarmbots Rely on Peers for Survival." Duke Pratt School of Engineering, February 29, 2016, https://pratt.duke.edu/about/news/engineered-swarmbots-rely-peers-survival.

European Commission, Directorate-General for Research. *Synthetic Biology: A NEST Pathfinder Initiative*, 2007, www.eurosfaire.prd.fr/7pc/doc/1182320848_5_nest_synthetic_080507.pdf.

Evans, Sam Weiss. "Synthetic Biology: Missing the Point." *Nature* 510, no. 7504 (June 2014).

Extance, Andy. "The First Gene on Earth May Have Been a Hybrid." *Scientific American*, June 22, 2020, www.scientificamerican.com/article/the-first-gene-on-earth-may-have-been-a-hybrid.

Farny, Natalie G. "A Vision for Teaching the Values of Synthetic Biology." *Trends in Biotechnology* 36, no. 11 (November 1, 2018): 1097–1100. https://doi.org/10.1016/j.tibtech.2018.07.019.

"FBI Laboratory Positions." Federal Bureau of Investigation, www.fbi.gov/services/laboratory/laboratory-positions.

Filosa, Gwen. "GMO Mosquitoes Have Landed in the Keys. Here's What You Need to Know." *Miami Herald*, May 3, 2021, www.miamiherald.com/news/local/community/florida-keys/article251031419.html.

Fisher, R. A. "The Use of Multiple Measurements in Taxonomic Problems." *Annals of Eugenics* 7, no. 2 (1936): 179–88. https://doi.org/10.1111/j.1469-1809.1936.tb02137.x.

———. "The Wave of Advance of Advantageous Genes." *Annals of Eugenics* 7, no. 4 (1937): 355–69. https://doi.org/10.1111/j.1469-1809.1937.tb02153.x.

Fralick, Michael, and Aaron S. Kesselheim. "The U.S. Insulin Crisis—Rationing a Lifesaving Medication Discovered in the 1920s." *New England Journal of Medicine* 381, no. 19 (November 7, 2019): 1793–95. https://doi.org/10.1056/NEJMp1909402.

French, H. *Midnight in Peking: How the Murder of a Young Englishwoman Haunted the Last Days of Old China*, rev. ed. New York: Penguin, 2012.

Friedman, Jared. "How Biotech Startup Funding Will Change in the Next 10 Years." YC Startup Library, n.d., www.ycombinator.com/library/4L-how-biotech-startup-funding-will-change-in-the-next-10-years.

Funk, Cary. "How Much the Public Knows About Science, and Why It Matters." *Scientific American*, April 9, 2019, https://blogs.scientificamerican.com/observations/how-much-the-public-knows-about-science-and-why-it-matters.

Gao, Huirong, Mark J. Gadlage, H. Renee Lafitte, Brian Lenderts, Meizhu Yang, Megan Schroder, Jeffry Farrell, et al. "Superior Field Performance of Waxy Corn Engineered Using CRISPR-Cas9." *Nature Biotechnology* 38, no. 579 (March 9, 2020).

"Genetics and Genomics Timeline: 1995." Genome News Network, www.genomenewsnetwork.org/resources/timeline/1995_Haemophilus.php.

"George Church." *Colbert Report*, season 9, episode 4, October 4, 2012 (video clip). Comedy Central, www.cc.com/video-clips/fkt99i/the-colbert-report-george-church.

"George Church" (oral history). National Human Genome Research Institute, National Institutes of Health, July 26, 2017, www.genome.gov/Multimedia/Transcripts/OralHistory/GeorgeChurch.pdf.

"German Research Bodies Draft Synthetic-Biology Plan." *Nature* 460, no. 563 (July 2009): 563, www.nature.com/articles/460563a.

Gilbert, C., and T. Ellis. "Biological Engineered Living Materials: Growing Functional Materials with Genetically Programmable Properties." *ACS Synthetic Biology* 8, no. 1 (2019).

Gostin, Lawrence O., Bruce M. Altevogt, and Andrew M. Pope. "Future Oversight of Recombinant DNA Research: Recommendations of an Institute of Medicine Committee." *JAMA* 311, no. 7 (February 19, 2014): 671–72. https://doi.org/10.1001/jama.2013.286312.

Gronvall, Gigi Kwik. "US Competitiveness in Synthetic Biology." *Health Security* 13, no. 6 (December 1, 2015): 378–89. https://doi.org/10.1089/hs.2015.0046.

Gross, Michael. "What Exactly Is Synthetic Biology?" *Current Biology* 21, no. 16 (August 23, 2011): R611–14. https://doi.org/10.1016/j.cub.2011.08.002.

Grushkin, Daniel. "The Rise and Fall of the Company That Was Going to Have Us All Using Biofuels." *Fast Company*, August 8, 2012, www.fastcompany.com/3000040/rise-and-fall-company-was-going-have-us-all-using-biofuels.

"Hacking DNA Sequences: Biosecurity Meets Cybersecurity." American Council on Science and Health, January 14, 2021, www.acsh.org/news/2021/01/14/hacking-dna-sequences-biosecurity-meets-cybersecurity-15273.

Hale, Piers J. "Monkeys into Men and Men into Monkeys: Chance and Contingency in the Evolution of Man, Mind and Morals in Charles Kingsley's Water Babies." *Journal of the History of Biology* 46, no. 4 (November 1, 2013): 551–97. https://doi.org/10.1007/s10739-012-9345-5.

Hall, Stephen S. "New Gene-Editing Techniques Could Transform Food Crops—or Die on the Vine." *Scientific American*, March 1, 2016, www.scientificamerican.com/article/new-gene-editing-techniques-could-transform-food-crops-or-die-on-the-vine. https://doi.org/10.1038/scientificamerican0316-56.

Harmon, Amy. "Golden Rice: Lifesaver?" *New York Times*, August 24, 2013, www.nytimes.com/2013/08/25/sunday-review/golden-rice-lifesaver.html.

———. "My Genome, Myself: Seeking Clues in DNA." *New York Times*, November 17, 2007, www.nytimes.com/2007/11/17/us/17dna.html.

———. "6 Billion Bits of Data About Me, Me, Me!" *New York Times*, June 3, 2007, www.nytimes.com/2007/06/03/weekinreview/03harm.html.

Harmon, Katherine. "Endangered Species Get Iced in Museum DNA Repository." *Scientific American*, July 8, 2009, www.scientificamerican.com/article/endangered-species-dna.

———. "Gene Sequencing Reveals the Dynamics of Ancient Epidemics." *Scientific American*, September 1, 2013, www.scientificamerican.com/article/gene-sequencing-reveals-the-dynamics-of-ancient-epidemics. https://doi.org/10.1038/scientificamerican0913-24b.

"He Jiankui's Gene Editing Experiment Ignored Other HIV Strains," Stat News, April 15, 2019, www.statnews.com/2019/04/15/jiankui-embryo-editing-ccr5.

Heinemann, Matthias, and Sven Panke. "Synthetic Biology: Putting Engineering into Bioengineering." In *Systems Biology and Synthetic Biology*, ed. Pengcheng Fu and Sven Panke, 387–409. Hoboken, NJ: John Wiley and Sons, 2009. https://doi.org/10.1002/9780470437988.ch11.

Herrera, Stephan. "Synthetic Biology Offers Alternative Pathways to Natural Products." *Nature Biotechnology* 23, no. 3 (March 1, 2005): 270–71. https://doi.org/10.1038/nbt0305-270.

"How Diplomacy Helped to End the Race to Sequence the Human Genome." *Nature* 582, no. 7813 (June 24, 2020): 460. https://doi.org/10.1038/d41586-020-01849-w.

"How Do Scientists Turn Genes on and off in Living Animals?" *Scientific American*, August 8, 2005, www.scientificamerican.com/article/how-do-scientists-turn-ge.

Ingbar, Sasha. "Japan's Population Is in Rapid Decline." National Public Radio, December 21, 2018, www.npr.org/2018/12/21/679103541/japans-population-is-in-rapid-decline.

Institute of Medicine, Committee on the Economics of Antimalarial Drugs. *Saving Lives, Buying Time: Economics of Malaria Drugs in an Age of Resistance*, eds. Kenneth J. Arrow, Claire Panosian, and Hellen Gelband. Washington, DC: National Academies Press, 2004.

Institute of Medicine, Committee to Study Decision Making, Division of Health Sciences Policy. *Biomedical Politics*, ed. Kathi E. Hanna. Washington, DC: National Academies Press, 1991.

Isaacs, Farren J., Daniel J. Dwyer, and James J. Collins. "RNA Synthetic Biology." *Nature Biotechnology* 24, no. 5 (May 2006): 545–54. https://doi.org/10.1038/nbt1208.

Jenkins, McKay. *Food Fight: GMOs and the Future of the American Diet*. New York: Penguin, 2018.

Jia, Jing, Yi-Liang Wei, Cui-Jiao Qin, Lan Hu, Li-Hua Wan, and Cai-Xia Li. "Developing a Novel Panel of Genome-Wide Ancestry Informative Markers for Bio-Geographical Ancestry Estimates." *Forensic Science International: Genetics* 8, no. 1 (January 2014): 187–94. https://doi.org/10.1016/j.fsigen.2013.09.004.

Jones, Richard. "The Question of Complexity." *Nature Nanotechnology* 3, no. 5 (May 2008): 245–46. https://doi.org/10.1038/nnano.2008.117.

Juhas, Mario, Leo Eberl, and George M. Church. "Essential Genes as Antimicrobial Targets and Cornerstones of Synthetic Biology." *Trends in Biotechnology* 30, no. 11 (November 1, 2012): 601–7. https://doi.org/10.1016/j.tibtech.2012.08.002.

Kania, Elsa B., and Wilson Vorndick. "Weaponizing Biotech: How China's Military Is Preparing for a 'New Domain of Warfare.'" Defense One, August 14, 2019, www.defenseone.com/ideas/2019/08/chinas-military-pursuing-biotech/159167.

Karp, David. "Most of America's Fruit Is Now Imported. Is That a Bad Thing?" *New York Times*, March 13, 2018, www.nytimes.com/2018/03/13/dining/fruit-vegetables-imports.html.

Keating, K. W., and E. M. Young. "Synthetic Biology for Bio-Derived Structural Materials." *Current Opinion in Chemical Engineering* 24, no. 107 (2019).

Keim, Brandon. "James Watson Suspended from Lab, But Not for Being a Sexist Hater of Fat People." *Wired*, October 2007, www.wired.com/2007/10/james-watson-su.

Kerlavage, Anthony R., Claire M. Fraser, and J. Craig Venter. "Muscarinic Cholinergic Receptor Structure: Molecular Biological Support for Subtypes." *Trends in Pharmacological Sciences* 8, no. 11 (November 1, 1987): 426–31. https://doi.org/10.1016/0165-6147(87)90230-6.

Kettenburg, Annika J., Jan Hanspach, David J. Abson, and Joern Fischer. "From Disagreements to Dialogue: Unpacking the Golden Rice Debate." *Sustainability Science* 13, no. 5 (2018): 1469–82. https://doi.org/10.1007/s11625-018-0577-y.

Kovelakuntla, Vamsi, and Anne S. Meyer. "Rethinking Sustainability Through Synthetic Biology." *Nature Chemical Biology*, May 10, 2021, 1–2. https://doi.org/10.1038/s41589-021-00804-8.

Kramer, Moritz. "Epidemiological Data from the NCoV-2019 Outbreak: Early Descriptions from Publicly Available Data." Virological, January 23, 2020, https://virological.org/t/epidemiological-data-from-the-ncov-2019-outbreak-early-descriptions-from-publicly-available-data/337.

Lander, Eric S. "Brave New Genome." *New England Journal of Medicine* 373, no. 1 (July 2, 2015): 5–8. https://doi.org/10.1056/NEJMp1506446.

Lane, Nick. *The Vital Question: Energy, Evolution, and the Origins of Complex Life*. New York: W. W. Norton, 2015.

Lavickova, Barbora, Nadanai Laohakunakorn, and Sebastian J. Maerkl. "A Partially Self-Regenerating Synthetic Cell." *Nature Communications* 11, no. 1 (December 11, 2020): 6340. https://doi.org/10.1038/s41467-020-20180-6.

Lentzos, Filippa. "How to Protect the World from Ultra-Targeted Biological Weapons." *Bulletin of the Atomic Scientists*, December 7, 2020, https://thebulletin.org/premium/2020-12/how-to-protect-the-world-from-ultra-targeted-biological-weapons.

Lin, F. K., S. Suggs, C. H. Lin, J. K. Browne, R. Smalling, J. C. Egrie, K. K. Chen, G. M. Fox, F. Martin, and Z. Stabinsky. "Cloning and Expression of the Human Erythropoietin Gene." *Proceedings of the National Academy of Sciences* 82, no. 22 (1985): 7580–84. https://doi.org/10.1073/pnas.82.22.7580.

Liu, Wusheng, and C. Neal Stewart. "Plant Synthetic Biology." *Trends in Plant Science* 20, no. 5 (May 1, 2015): 309–17. https://doi.org/10.1016/j.tplants.2015.02.004.

Lynas, Mark. "Anti-GMO Activists Lie About Attack on Rice Crop (and About So Many Other Things)." Slate, August 26, 2013, https://slate.com/technology/2013/08/golden-rice-attack-in-philippines-anti-gmo-activists-lie-about-protest-and-safety.html.

Macilwain, Colin. "World Leaders Heap Praise on Human Genome Landmark." *Nature* 405, no. 6790 (June 1, 2000): 983. https://doi.org/10.1038/35016696.

Malech, Harry L. "Treatment by CRISPR-Cas9 Gene Editing—A Proof of Principle." *New England Journal of Medicine* 384, no. 3 (January 21, 2021): 286–87. https://doi.org/10.1056/NEJMe2034624.

Mali, Prashant, Luhan Yang, Kevin M. Esvelt, John Aach, Marc Guell, James E. DiCarlo, Julie E. Norville, and George M. Church. "RNA-Guided Human Genome Engineering via Cas9." *Science* 339, no. 6121 (February 15, 2013): 823–26. https://doi.org/10.1126/science.1232033.

Marner, Wesley D. "Practical Application of Synthetic Biology Principles." *Biotechnology Journal* 4, no. 10 (2009): 1406–19. https://doi.org/10.1002/biot.200900167.

Maxson Jones, Kathryn, Rachel A. Ankeny, and Robert Cook-Deegan. "The Bermuda Triangle: The Pragmatics, Policies, and Principles for Data Sharing in the History of the Human Genome Project." *Journal of the History of Biology* 51, no. 4 (December 1, 2018): 693–805. https://doi.org/10.1007/s10739-018-9538-7.

Menz, J., D. Modrzejewski, F. Hartung, R. Wilhelm, and T. Sprink. "Genome Edited Crops Touch the Market: A View on the Global Development and Regulatory Environment." *Frontiers in Plant Science* 11, no. 586027 (2020).

Metzl, Jamie. *Hacking Darwin: Genetic Engineering and the Future of Humanity*. Naperville, IL: Sourcebooks, 2019.

Mitka, Mike. "Synthetic Cells." *JAMA* 304, no. 2 (July 14, 2010): 148. https://doi.org/10.1001/jama.2010.879.

"Modernizing the Regulatory Framework for Agricultural Biotechnology Products." Federal Register, June 14, 2019, www.federalregister.gov/documents/2019/06/14/2019-12802/modernizing-the-regulatory-framework-for-agricultural-biotechnology-products.

Molteni, Megan. "California Could Be First to Mandate Biosecurity for Mail-Order DNA." Stat News, May 20, 2021, www.statnews.com/2021/05/20/california-could-become-first-state-to-mandate-biosecurity-screening-by-mail-order-dna-companies.

Moore, James. "Deconstructing Darwinism: The Politics of Evolution in the 1860s." *Journal of the History of Biology* 24, no. 3 (September 1, 1991): 353–408. https://doi.org/10.1007/BF00156318.

Mora, Camilo, Chelsie W. W. Counsell, Coral R. Bielecki, and Leo V. Louis. "Twenty-Seven Ways a Heat Wave Can Kill You: Deadly Heat in the Era of Climate Change." *Circulation: Cardiovascular Quality and Outcomes* 10, no. 11 (November 1, 2017), https://doi.org/10.1161/CIRCOUTCOMES.117.004233.

Morowitz, Harold J. "Thermodynamics of Pizza." *Hospital Practice* 19, no. 6 (June 1, 1984): 255–58. https://doi.org/10.1080/21548331.1984.11702854.

Mukherjee, Siddhartha. *The Gene: An Intimate History*. New York: Scribner, 2016.

Müller, K. M., and K. M. Arndt. "Standardization in Synthetic Biology." *Methods in Molecular Biology* 813 (2012): 23–43.

Musk, Elon. "Making Humans a Multi-Planetary Species." *New Space* 5, no. 2 (June 1, 2017): 46–61. https://doi.org/10.1089/space.2017.29009.emu.

National Academies of Sciences, Engineering, and Medicine. *Biodefense in the Age of Synthetic Biology*. Washington, DC: National Academies Press, 2018. https://doi.org/10.17226/24890.

———. *The Current Biotechnology Regulatory System: Preparing for Future Products of Biotechnology*. Washington, DC: National Academies Press, 2017.

———. *Safeguarding the Bioeconomy*. Washington, DC: National Academies Press, 2020. https://doi.org/10.17226/25525.

Nielsen, Jens, and Jay D. Keasling. "Engineering Cellular Metabolism." *Cell* 164, no. 6 (March 10, 2016): 1185–97. https://doi.org/10.1016/j.cell.2016.02.004.

"No More Needles! Using Microbiome and Synthetic Biology Advances to Better Treat Type 1 Diabetes." J. Craig Venter Institute, March 25, 2019, www.jcvi.org/blog/no-more-needles-using-microbiome-and-synthetic-biology-advances-better-treat-type-1-diabetes.

O'Neill, Helen C., and Jacques Cohen. "Live Births Following Genome Editing in Human Embryos: A Call for Clarity, Self-Control and Regulation." Reproductive BioMedicine Online 38, no. 2 (February 1, 2019): 131–32. https://doi.org/10.1016/j.rbmo.2018.12.003.

Ossola, Alexandra. "Scientists Build a Living Cell with Minimum Viable Number of Genes." *Popular Science*, March 24, 2016, www.popsci.com/scientists-create-living-cell-with-minimum-number-genes.

"Park History." Asilomar Conference Grounds, www.visitasilomar.com/discover/park-history.

"Parties to the Cartagena Protocol and Its Supplementary Protocol on Liability and Redress." Convention on Biological Diversity, https://bch.cbd.int/protocol/parties.

Patterson, Andrea. "Germs and Jim Crow: The Impact of Microbiology on Public Health Policies in Progressive Era American South." *Journal of the History of Biology* 42, no. 3 (October 29, 2008): 529. https://doi.org/10.1007/s10739-008-9164-x.

People's Republic of China, State Council. Made in China 2025. July 2015.

———. Notice on the Publication of the National 13th Five-Year Plan for S&T Innovation. July 2016.

Pinker, Steven. "My Genome, My Self." *New York Times*, January 7, 2009, www.nytimes.com/2009/01/11/magazine/11Genome-t.html.

"Polynucleotide Synthesizer Model 280, Solid Phase Microprocessor Controller Model 100B." National Museum of American History, https://americanhistory.si.edu/collections/search/object/nmah_1451158.

"President Clinton Announces the Completion of the First Survey of the Entire Human Genome." White House Press Release, June 25, 2000. Human Genome Project Information Archive, 1990–2003, https://web.ornl.gov/sci/techresources/Human_Genome/project/clinton1.shtml.

"Press Briefing by Dr. Neal Lane, Assistant to the President for Science and Technology; Dr. Frances Collins, Director of the National Human Genome Research Institute; Dr. Craig Venter, President and Chief Scientific Officer, Celera Genomics Corporation; and Dr. Ari Patrinos, Associate Director for Biological and Environmental Research, Department of Energy, on the Completion of the First Survey of the Entire Human Genome." White House Press Release, June 26, 2000. Human Genome Project Information Archive, 1990–2003, https://web.ornl.gov/sci/techresources/Human_Genome/project/clinton3.shtml.

Puzis, Rami, Dor Farbiash, Oleg Brodt, Yuval Elovici, and Dov Greenbaum. "Increased Cyber-Biosecurity for DNA Synthesis." *Nature Biotechnology* 38, no. 12 (December 2020): 1379–81. https://doi.org/10.1038/s41587-020-00761-y.

Race, Tim. "New Economy: There's Gold in Human DNA, and He Who Maps It First Stands to Win on the Scientific, Software and Business Fronts." *New York Times*, June 19, 2000, www.nytimes.com/2000/06/19/business/new-economy-there-s-gold-human-dna-he-who-maps-it-first-stands-win-scientific.html.

"Reading the Book of Life: White House Remarks on Decoding of Genome." *New York Times*, June 27, 2000, www.nytimes.com/2000/06/27/science/reading-the-book-of-life-white-house-remarks-on-decoding-of-genome.html.

Reardon, Sara. "US Government Lifts Ban on Risky Pathogen Research." *Nature* 553, no. 7686 (December 19, 2017): 11. https://doi.org/10.1038/d41586-017-08837-7.

Regis, Ed. "Golden Rice Could Save Children. Until Now, Governments Have Barred It." *Washington Post*, November 11, 2019, www.washingtonpost.com/opinions/2019/11/11/golden-rice-long-an-anti-gmo-target-may-finally-get-chance-help-children.

———. "The True Story of the Genetically Modified Superfood That Almost Saved Millions." *Foreign Policy*, October 17, 2019, https://foreignpolicy.com/2019/10/17/golden-rice-genetically-modified-superfood-almost-saved-millions.

Remington, Karin A., Karla Heidelberg, and J. Craig Venter. "Taking Metagenomic Studies in Context." *Trends in Microbiology* 13, no. 9 (September 1, 2005): 404. https://doi.org/10.1016/j.tim.2005.07.001.

Rich, Nathaniel. "The Mammoth Cometh." *New York Times*, February 27, 2014, www.nytimes.com/2014/03/02/magazine/the-mammoth-cometh.html.

Ro, D. K., E. Paradise, M. Ouellet, K. J. Fisher, K. L. Newman, J. M. Ndungu, K. A. Ho, et al. "Production of the Antimalarial Drug Precursor Artemisinic Acid in Engineered Yeast." *Nature* 440, no. 7086 (2006): 940–43. https://doi.org/10.1038/nature04640.

Robbins, Rebecca. "A Genomics Pioneer Is Selling a Full DNA Analysis for $1,400. Is It Worth It?" Stat News, March 21, 2017, www.statnews.com/2017/03/21/craig-venter-sequence-genome.

———. "Judge Dismisses Lawsuit Accusing Craig Venter of Stealing Trade Secrets." Stat News, December 19, 2018, www.statnews.com/2018/12/19/judge-dismisses-lawsuit-accusing-craig-venter-of-stealing-trade-secrets.

Roosth, Sophia. *Synthetic—How Life Got Made*. Chicago: University of Chicago Press, 2017.

Rutjens, Bastiaan. "What Makes People Distrust Science? Surprisingly, Not Politics." Aeon, May 28, 2018, https://aeon.co/ideas/what-makes-people-distrust-science-surprisingly-not-politics.

Salem, Iman, Amy Ramser, Nancy Isham, and Mahmoud A. Ghannoum. "The Gut Microbiome as a Major Regulator of the Gut-Skin Axis." *Frontiers in Microbiology* 9 (July 10, 2018). https://doi.org/10.3389/fmicb.2018.01459.

Scarfuto, Jessica. "Do You Trust Science? These Five Factors Play a Big Role." *Science*, February 16, 2020, www.sciencemag.org/news/2020/02/do-you-trust-science-these-five-factors-play-big-role.

Schmidt, Markus, Malcolm Dando, and Anna Deplazes. "Dealing with the Outer Reaches of Synthetic Biology Biosafety, Biosecurity, IPR, and Ethical Challenges of Chemical Synthetic Biology." In *Chemical Synthetic Biology*, ed. P. L. Luisi and C. Chiarabelli, 321–42. New York: John Wiley and Sons, 2011. https://doi.org/10.1002/9780470977873.ch13.

Scudellari, Megan. "Self-Destructing Mosquitoes and Sterilized Rodents: The Promise of Gene Drives." *Nature* 571, no. 7764 (July 9, 2019): 160–62. https://doi.org/10.1038/d41586-019-02087-5.

Selberg, John, Marcella Gomez, and Marco Rolandi. "The Potential for Convergence Between Synthetic Biology and Bioelectronics." *Cell Systems* 7, no. 3 (September 26, 2018): 231–44. https://doi.org/10.1016/j.cels.2018.08.007.

Simon, Matt. "Climate Change Is Turning Cities into Ovens." *Wired*, January 7, 2021, www.wired.com/story/climate-change-is-turning-cities-into-ovens.

Skerker, Jeffrey M., Julius B. Lucks, and Adam P. Arkin. "Evolution, Ecology and the Engineered Organism: Lessons for Synthetic Biology." *Genome Biology* 10, no. 11 (November 30, 2009): 114. https://doi.org/10.1186/gb-2009-10-11-114.

Sprinzak, David, and Michael B. Elowitz. "Reconstruction of Genetic Circuits." *Nature* 438, no. 7067 (November 2005): 443–48. https://doi.org/10.1038/nature04335.

Telenti, Amalio, Brad A. Perkins, and J. Craig Venter. "Dynamics of an Aging Genome." *Cell Metabolism* 23, no. 6 (June 14, 2016): 949–50. https://doi.org/10.1016/j.cmet.2016.06.002.

Topol, Eric. "A Deep and Intimate Inquiry of Genes." *Cell* 165, no. 6 (June 2, 2016): 1299–1300. https://doi.org/10.1016/j.cell.2016.05.065.

US Department of Defense. "Summary of the 2018 National Defense Strategy of the United States of America: Sharpening the American Military's Competitive Edge." 2018, https://dod.defense.gov/Portals/1/Documents/pubs/2018-National-Defense-Strategy-Summary.pdf.

US Department of Health and Human Services, Office of the Assistant Secretary for Preparedness and Response (ASPR). "National Health Security Strategy, 2019–2222." ASPR, 2019, www.phe.gov/Preparedness/planning/authority/nhss/Pages/default.aspx.

US Department of Health and Human Services and US Department of Energy. "Understanding Our Genetic Inheritance. The Human Genome Project: The First Five Years, FY 1991–1995." DOE/ER-0452P, April 1990, https://web.ornl.gov/sci/techresources/Human_Genome/project/5yrplan/firstfiveyears.pdf.

US Department of State and US Agency for International Development, "Joint Strategic Plan FY 2018–2022," February 2018, www.state.gov/wp-content/uploads/2018/12/Joint-Strategic-Plan-FY-2018-2022.pdf.

Venter, J. Craig. *Life at the Speed of Light*. New York: Viking, 2013.

Venter, J. Craig, Mark D. Adams, Antonia Martin-Gallardo, W. Richard McCombie, and Chris Fields. "Genome Sequence Analysis: Scientific Objectives and Practical Strategies." *Trends in Biotechnology* 10 (January 1, 1992): 8–11. https://doi.org/10.1016/0167-7799(92)90158-R.

Venter, J. Craig, and Claire M. Fraser. "The Structure of α- and β-Adrenergic Receptors." *Trends in Pharmacological Sciences* 4 (January 1, 1983): 256–58. https://doi.org/10.1016/0165-6147(83)90390-5.

Vinge, V. "The Coming Technological Singularity: How to Survive in the Post-Human Era." In *Vision-21: Interdisciplinary Science and Engineering in the Era of Cyberspace*, NASA Conference Publication 10129, 1993, 11–22, http://ntrs.nasa.gov/archive/nasa/casi.ntrs.nasa.gov/19940022855_1994022855.pdf.

Waltz, Emily. "Gene-Edited CRISPR Mushroom Escapes US Regulation: Nature News and Comment." *Nature* 532, no. 293 (2016). www.nature.com/news/gene-edited-crispr-mushroom-escapes-us-regulation-1.19754.

Webb, Amy. "CRISPR Makes It Clear: The US Needs a Biology Strategy, and Fast." *Wired*, May 11, 2017, www.wired.com/2017/05/crispr-makes-clear-us-needs-biology-strategy-fast.

Wee, Sui-Lee. "China Uses DNA to Track Its People, with the Help of American Expertise." *New York Times*, February 21, 2019, www.nytimes.com/2019/02/21/business/china-xinjiang-uighur-dna-thermo-fisher.html.

Weiss, Robin A. "Robert Koch: The Grandfather of Cloning?" *Cell* 123, no. 4 (November 18, 2005): 539–42. https://doi.org/10.1016/j.cell.2005.11.001.

Weiss, Ron, Joseph Jacobson, Paul Modrich, Jim Collins, George Church, Christina Smolke, Drew Endy, David Baker, and Jay Keasling. "Engineering Life: Building a FAB for Biology." *Scientific American*, June 2006, www.scientificamerican.com/article/engineering-life-building.

Weiss, Sheila Faith. "Human Genetics and Politics as Mutually Beneficial Resources: The Case of the Kaiser Wilhelm Institute for Anthropology, Human Heredity and Eugenics During the Third Reich." *Journal of the History of Biology* 39, no. 1 (March 1, 2006): 41–88. https://doi.org/10.1007/s10739-005-6532-7.

White House, National Biodefense Strategy. Washington, DC: White House, 2018.

White House. "White House Precision Medicine Initiative." https://obamawhitehouse.archives.gov/node/333101.

Wickiser, J. Kenneth, Kevin J. O'Donovan, Michael Washington, Stephen Hummel, and F. John Burpo. "Engineered Pathogens and Unnatural Biological Weapons: The Future Threat of Synthetic Biology," *CTC Sentinel* 13, no. 8 (August 31, 2020): 1–7, https://ctc.usma.edu/engineered-pathogens-and-unnatural-biological-weapons-the-future-threat-of-synthetic-biology.

Wong, Pak Chung, Kwong-kwok Wong, and Harlan Foote. "Organic Data Memory Using the DNA Approach." *Communications of the ACM* 46, no. 1 (January 2003): 95–98. https://doi.org/10.1145/602421.602426.

Wood, Sara, Jeremiah A. Henning, Luoying Chen, Taylor McKibben, Michael L. Smith, Marjorie Weber, Ash Zemenick, and Cissy J. Ballen. "A Scientist Like Me: Demographic Analysis of Biology Textbooks Reveals Both Progress and Long-Term Lags." *Proceedings of the Royal Society B: Biological Sciences* 287, no. 1929 (June 24, 2020): 20200877. https://doi.org/10.1098/rspb.2020.0877.

Woolfson, Adrian. *Life Without Genes*. New York: HarperCollins, 2000.

Wu, Katherine J. "Scientific Journals Commit to Diversity but Lack the Data." *New York Times*, October 30, 2020, www.nytimes.com/2020/10/30/science/diversity-science-journals.html.

Wurtzel, Eleanore T., Claudia E. Vickers, Andrew D. Hanson, A. Harvey Millar, Mark Cooper, Kai P. Voss-Fels, Pablo I. Nikel, and Tobias J. Erb. "Revolutionizing Agriculture with Synthetic Biology." *Nature Plants* 5, no. 12 (December 2019): 1207–10. https://doi.org/10.1038/s41477-019-0539-0.

Yamey, Gavin. "Scientists Unveil First Draft of Human Genome." *British Medical Journal* 321, no. 7252 (July 1, 2000): 7.

Yang, Annie, Zhou Zhu, Philipp Kapranov, Frank McKeon, George M. Church, Thomas R. Gingeras, and Kevin Struhl. "Relationships Between P63 Binding, DNA Sequence, Transcription Activity, and Biological Function in Human Cells." *Molecular Cell* 24, no. 4 (November 17, 2006): 593–602. https://doi.org/10.1016/j.molcel.2006.10.018.

Yetisen, Ali K., Joe Davis, Ahmet F. Coskun, George M. Church, and Seok Hyun Yun. "Bioart." *Trends in Biotechnology* 33, no. 12 (December 1, 2015): 724–34. https://doi.org/10.1016/j.tibtech.2015.09.011.

Zayner, Josiah. "How to Genetically Engineer a Human in Your Garage. Part III—The First Round of Experiments." Science, Art, Beauty, February 15, 2017, www.josiahzayner.com/2017/02/how-to-genetically-engineer-human-part.html.

Zimmer, Carl. "James Joyce's Words Come to Life, and Are Promptly Desecrated." *Discover*, May 21, 2010, www.discovermagazine.com/planet-earth/james-joyces-words-come-to-life-and-are-promptly-desecrated.

# *INDEX*

ELENA SEIBERT

**Amy Webb** is the author of several popular books about the futures of science and technology, including *The Big Nine: How the Tech Titans and Their Thinking Machines Could Warp Humanity*, which was longlisted for the *Financial Times* and McKinsey Business Book of the Year Award, shortlisted for the Thinkers50 Digital Thinking Award, and won the 2020 Gold Axiom Medal for the best book about business and technology. An earlier book, *The Signals Are Talking: Why Today's Fringe Is Tomorrow's Mainstream*, won the Thinkers50 Radar Award and the 2017 Gold Axiom Medal for the best book about business and technology. It was also selected as one of *Fast Company*'s Best Business Books of 2016 and one of Amazon's Best Books of December 2016. Webb advises CEOs of some of the world's most admired companies as well as three-star admirals and generals and the senior leadership of central banks and intergovernmental organizations. Founder of the Future Today Institute, a leading foresight and strategy firm that helps leaders and their organizations prepare for complex futures, Amy pioneered a data-driven, technology-led foresight methodology that is now used within hundreds of organizations.

She is a professor of strategic foresight at the New York University Stern School of Business, where she developed and teaches the MBA course on strategic foresight; a Visiting Fellow at Oxford University's Säid Business School; a Nonresident Senior Fellow at the Atlantic Council's GeoTech Center; a Fellow in the US-Japan Leadership Program; and a Foresight Fellow in the US Government Accountability Office's Center for Strategic Foresight. She was elected a life member to the Council on Foreign Relations and is a member of the Bretton Woods Committee, and she is a member of the World Economic Forum, where she serves on the Global Future Council on Media, Entertainment and Culture and the Stewardship Board of the Forum's Platform for Shaping the Future of Media, Entertainment and Culture. She was also a Visiting Nieman Fellow at Harvard University, where her research received a national Sigma Delta

Chi award, and a delegate on the former US-Russia Bilateral Presidential Commission, where she worked on the future of technology, media, and international diplomacy.

A lifelong science fiction fan, Amy collaborates closely with Hollywood writers and producers on films, TV shows, and commercials about science, technology, and the future. She was named by *Forbes* as one of the "Women Changing the World," selected by the BBC for its 100 Women of 2020 list of "inspiring and influential women from around the world," and named to the Thinkers50 Radar list "of the 30 management thinkers most likely to shape the future of how organizations are managed and led." She is based in New York City.

HUMANE GENOMICS

**Andrew Hessel** is a geneticist, entrepreneur, and science communicator exploring the front lines of digital biology with a focus on complete genome synthesis. A former researcher with Amgen and Autodesk, he is a cofounder of Humane Genomics, Inc., a New York City–based biotechnology company specializing in designer artificial viruses that target cancer cells. He also cofounded Genome Project–Write (GPW), the international scientific effort advancing the design, construction, and testing of large genomes, including the human genome, and serves as its chairman. His current projects involve biotechnology and blockchain technologies, online biofoundries, and closed sustainable ecosystems. He is based in San Francisco, California.

B B S

PublicAffairs is a publishing house founded in 1997. It is a tribute to the standards, values, and flair of three persons who have served as mentors to countless reporters, writers, editors, and book people of all kinds, including me.

I. F. STONE, proprietor of *I. F. Stone's Weekly*, combined a commitment to the First Amendment with entrepreneurial zeal and reporting skill and became one of the great independent journalists in American history. At the age of eighty, Izzy published *The Trial of Socrates*, which was a national bestseller. He wrote the book after he taught himself ancient Greek.

BENJAMIN C. BRADLEE was for nearly thirty years the charismatic editorial leader of *The Washington Post*. It was Ben who gave the *Post* the range and courage to pursue such historic issues as Watergate. He supported his reporters with a tenacity that made them fearless and it is no accident that so many became authors of influential, best-selling books.

ROBERT L. BERNSTEIN, the chief executive of Random House for more than a quarter century, guided one of the nation's premier publishing houses. Bob was personally responsible for many books of political dissent and argument that challenged tyranny around the globe. He is also the founder and longtime chair of Human Rights Watch, one of the most respected human rights organizations in the world.

• • •

For fifty years, the banner of Public Affairs Press was carried by its owner Morris B. Schnapper, who published Gandhi, Nasser, Toynbee, Truman, and about 1,500 other authors. In 1983, Schnapper was described by *The Washington Post* as "a redoubtable gadfly." His legacy will endure in the books to come.

Peter Osnos, *Founder*